Lecture Notes in Physics

For information about Vols. 1–151, please contact your bookseller or Springer-Verlag.

Vol. 152: Physics of Narrow Gap Semiconductors. Proceedings, 1981. Edited by E. Gornik, H. Heinrich and L. Palmetshofer. XIII, 485 pages. 1982.

Vol. 153: Mathematical Problems in Theoretical Physics. Proceedings, 1981. Edited by R. Schrader, R. Seiler, and D.A. Uhlenbrock. XII, 429 pages. 1982.

Vol. 154: Macroscopic Properties of Disordered Media. Proceedings, 1981. Edited by R. Burridge, S. Childress, and G. Papanicolaou. VII, 307 pages. 1982.

Vol. 155: Quantum Optics. Proceedings, 1981. Edited by C.A. Engelbrecht. VIII, 329 pages. 1982.

Vol. 156: Resonances in Heavy Ion Reactions. Proceedings, 1981. Edited by K.A. Eberhard. XII, 448 pages. 1982.

Vol. 157: P. Niyogi, Integral Equation Method in Transonic Flow. XI, 189 pages. 1982.

Vol. 158: Dynamics of Nuclear Fission and Related Collective Phenomena. Proceedings, 1981. Edited by P. David, T. Mayer-Kuckuk, and A. van der Woude. X, 462 pages. 1982.

Vol. 159: E. Seiler, Gauge Theories as a Problem of Constructive Quantum Field Theory and Statistical Mechanics. V, 192 pages. 1982.

Vol. 160: Unified Theories of Elementary Particles. Critical Assessment and Prospects. Proceedings, 1981. Edited by P. Breitenlohner and H.P. Dürr. VI, 217 pages. 1982.

Vol. 161: Interacting Bosons in Nuclei. Proceedings, 1981. Edited by J.S. Dehesa, J.M.G. Gomez, and J. Ros. V, 209 pages. 1982.

Vol. 162: Relativistic Action at a Distance: Classical and Quantum Aspects. Proceedings, 1981. Edited by J. Llosa. X, 263 pages. 1982.

Vol. 163: J.S. Darrozes, C. Francois, Mécanique des Fluides Incompressibles. XIX, 459 pages. 1982.

Vol. 164: Stability of Thermodynamic Systems. Proceedings, 1981. Edited by J. Casas-Vázquez and G. Lebon. VII, 321 pages. 1982.

Vol. 165: N. Mukunda, H. van Dam, L.C. Biedenharn, Relativistic Models of Extended Hadrons Obeying a Mass-Spin Trajectory Constraint. Edited by A. Böhm and J.D. Dollard. VI, 163 pages. 1982.

Vol. 166: Computer Simulation of Solids. Edited by C.R.A. Catlow and W.C. Mackrodt. XII, 320 pages. 1982.

Vol. 167: G. Fieck, Symmetry of Polycentric Systems. VI, 137 pages, 1982.

Vol. 168: Heavy-Ion Collisions. Proceedings, 1982. Edited by G. Madurga and M. Lozano. VI, 429 pages. 1982.

Vol. 169: K. Sundermeyer, Constrained Dynamics. IV, 318 pages. 1982.

Vol. 170: Eighth International Conference on Numerical Methods in Fluid Dynamics. Proceedings, 1982. Edited by E. Krause. X, 569 pages. 1982.

Vol. 171: Time-Dependent Hartree-Fock and Beyond. Proceedings, 1982. Edited by K. Goeke and P.-G. Reinhard. VIII, 426 pages. 1982.

Vol. 172: Ionic Liquids, Molten Salts and Polyelectrolytes. Proceedings, 1982. Edited by K.-H. Bennemann, F. Brouers, and D. Quitmann. VII, 253 pages. 1982.

Vol. 173: Stochastic Processes in Quantum Theory and Statistical Physics. Proceedings, 1981. Edited by S. Albeverio, Ph. Combe, and M. Sirugue-Collin. VIII, 337 pages. 1982.

Vol. 174: A. Kadić, D.G.B. Edelen, A Gauge Theory of Dislocations and Disclinations. VII, 290 pages. 1983.

Vol. 175: Defect Complexes in Semiconductor Structures. Proceedings, 1982. Edited by J. Giber, F. Beleznay, J.C. Szép, and J. László. VI, 308 pages. 1983.

Vol. 176: Gauge Theory and Gravitation. Proceedings, 1982. Edited by K. Kikkawa, N. Nakanishi, and H. Nariai. X, 316 pages. 1983.

Vol. 177: Application of High Magnetic Fields in Semiconductor Physics. Proceedings, 1982. Edited by G. Landwehr. XII, 552 pages. 1983.

Vol. 178: Detectors in Heavy-Ion Reactions. Proceedings, 1982. Edited by W. von Oertzen. VIII, 258 pages. 1983.

Vol. 179: Dynamical Systems and Chaos. Proceedings, 1982. Edited by L. Garrido. XIV, 298 pages. 1983.

Vol. 180: Group Theoretical Methods in Physics. Proceedings, 1982. Edited by M. Serdaroğlu and E. İnönü. XI, 569 pages. 1983.

Vol. 181: Gauge Theories of the Eighties. Proceedings, 1982. Edited by R. Raitio and J. Lindfors. V, 644 pages. 1983.

Vol. 182: Laser Physics. Proceedings, 1983. Edited by J.D. Harvey and D.F. Walls. V, 263 pages. 1983.

Vol. 183: J.D. Gunton, M. Droz, Introduction to the Theory of Metastable and Unstable States. VI, 140 pages. 1983.

Vol. 184: Stochastic Processes – Formalism and Applications. Proceedings, 1982. Edited by G.S. Agarwal and S. Dattagupta. VI, 324 pages. 1983.

Vol. 185: H.N. Shirer, R. Wells, Mathematical Structure of the Singularities at the Transitions between Steady States in Hydrodynamic Systems. XI, 276 pages. 1983.

Vol. 186: Critical Phenomena. Proceedings, 1982. Edited by F.J.W. Hahne. VII, 353 pages. 1983.

Vol. 187: Density Functional Theory. Edited by J. Keller and J.L. Gázquez. V, 301 pages. 1983.

Vol. 188: A.P. Balachandran, G. Marmo, B.-S. Skagerstam, A. Stern, Gauge Symmetries and Fibre Bundles. IV, 140 pages. 1983.

Vol. 189: Nonlinear Phenomena. Proceedings, 1982. Edited by K.B. Wolf. XII, 453 pages. 1983.

Vol. 190: K. Kraus, States, Effects, and Operations. Edited by A. Böhm, J.W. Dollard and W.H. Wootters. IX, 151 pages. 1983.

Vol. 191: Photon Photon Collisions. Proceedings, 1983. Edited by Ch. Berger. V, 417 pages. 1983.

Vol. 192: Heidelberg Colloquium on Spin Glasses. Proceedings, 1983. Edited by J.L. van Hemmen and I. Morgenstern. VII, 356 pages. 1983.

Vol. 193: Cool Stars, Stellar Systems, and the Sun. Proceedings, 1983. Edited by S.L. Balliunas and L. Hartmann. VII, 364 pages. 1984.

Vol. 194: P. Pascual, R. Tarrach, QCD: Renormalization for the Practitioner. V, 277 pages. 1984.

Lecture Notes in Physics

Edited by H. Araki, Kyoto, J. Ehlers, München, K. Hepp, Zürich
R. Kippenhahn, München, H. A. Weidenmüller, Heidelberg
and J. Zittartz, Köln

222

A. García
P. Kielanowski

The Beta Decay of Hyperons

With Foreword and Addendum by A. Bohm

Lectures in Mathematics and Physics
at the University of Texas at Austin
Edited by A. Bohm

Springer-Verlag
Berlin Heidelberg GmbH

Authors

Augusto García
Departamento de Física, Centro de Investigación y de
Estudios Avanzados del Instituto Politécnico Nacional
Apartado Postal 14-740, 07000 México, D.F., México

Also at Escuela Supererior de Física y Mathemáticas (IPN),
México, D.F.

Piotr Kielanowski
Center for Particle Theory and the Physics Department
The University of Texas, Austin, TX 78712, USA

and

Departamento de Física, Centro de Investigación y de
Estudios Avanzados del Instituto Politécnico Nacional
Apartado Postal 14-740, 07000 México, D.F., México

On leave of absence from the University of Warsaw, Poland

Editor

Arno Bohm
Physics Department, The University of Texas
Austin, TX 78712, USA

ISBN 978-3-540-15184-5 ISBN 978-3-540-39263-7 (eBook)
DOI 10.1007/978-3-540-39263-7

Originally published by Springer-Verlag Berlin Heidelberg New York Tokyo in 1985

2153/3140-543210

Foreword

The purpose of the Texas lecture notes is to inform graduate students and "non-specialists" about recent developments in various areas of mathematics and physics. This volume is the result of a series of lectures delivered at the University of Texas at Austin during the Spring Semester, 1983. Besides the lecture notes, in Appendix 1 are found complete numerical formulas for transition rates, correlation coefficients, and asymmetries for all semileptonic decays, regardless of whether the parameter has been determined experimentally.

These notes are valuable to a student because they give a descriptive overview of the entire subject. In addition, they provide a lesson on how to perform a comparison between theoretical models and experimental numbers. Questions concerning the necessary degree of agreement between theory and experiment and the importance of theoretical corrections are emphasized. But these notes are also useful to the specialist working in the area because they give explicit expressions and numerical tables for the observable quantities.

When physics advances into a new domain of understanding, it can do this in two ways 1) it can proceed to new regions for the observables by, say, going to higher energies; 2) it can stay in the old region of energy and increase the precision. A good example is QED which has been explored by increasing the energy for electron scattering. However, the most useful experimental data came from experiments in the old domain of atomic physics which were performed with much higher resolution. Examples are the Lamb shift and the electron's anomalous magnetic moment. Hyperon semileptonic decay parameters and hyperon magnetic moments can now be measured to such a high accuracy that they can reveal fine structure effects which may be the remnants of major structures in a new domain of particle physics. Precision experiments with the "old" baryons may thus give us some insight into the physics of the "new" particles. The detailed formulas reported in this volume will be of great practical value in analyzing these precision experiments.

Although there are many details in the area of hyperon decays which are still untested, at the present time one may be tempted to accept the standard theoretical ideas, leave this area behind, and rush into the new domain of particle physics at higher energies. Unfortunately there is one experimental quantity, the electron asymmetry in $\Sigma^- \to ne\nu$, that has consistently shown deviations from the standard prediction. Thus there is also some excitement left in this old area.

A. Bohm
Editor

Preface

The motivation for writing these lecture notes can be described best with an anecdote. George Uhlenbeck spent the summer of 1971 visiting our Physics Deparment in Mexico. For two months he lectured a superb course in statistical mechanics. When he was done with it, he remarked he still had time for one more lecture and he asked his audience if someone wished that he covered a special topic. Somebody had the good idea to ask him to give us a first-hand version about the discovery of spin. He agreed and he scheduled that lecture for an afternoon a couple of days later.

When the time came, we were meeting in the classroom that had been regularly used for his course, but Uhlenbeck proposed we should go to the garden and sit on the grass, for it was such a lovely afternoon. A few minutes later, we were all waiting in the garden for him to come. Apparently, he was following the group towards the garden, but somehow he went elsewhere. He kept us waiting for some ten minutes until he finally showed up. He apologized for the delay saying that he needed a reference and he had gone to look for it, but on his way out of the library something happened to him. He said: "I have just realized what is wrong with high-energy physics". And then he explained:

"In my time, when spin was discovered, we had two thick volumes of experimental data (in saying this, he showed his hands with the thumbs separated from the other fingers as if he were holding one volume three inches thick in each hand). As I was coming out of the library, I passed by a shelf where the Reviews of Particle Data are collected and I noticed the latest issue... It is so thin! (and in saying this, he showed one hand with the thumb and forefingers streched and separated by about half an inch)... That is what is wrong with high-energy physics!"

Needless to say that his account emphasized the role those two thick volumes played in the discovery of spin.

Table of Contents

Chapter 1. Introduction

Almost[1] 50 years ago Fermi introduced the theory of β decay —the V-theory— incorporating the elusive neutrino proposed by Pauli. Next, 35 years ago Gamow and Teller showed that an A-Theory was necessary to explain other forms of nuclear β decay. Both theories were fused into the V-A Theory by Sudarshan and Marshak and also independently by Feynman and Gell-Mann, 26 years ago, motivated by the discovery of parity violation in weak interactions by Lee and Yang (theoretically) and Wu and Telegdi (experimentally). The detection of a β decay mode of strangeness carrying mesons and hyperons and the discovery of the SU(3) symmetry of hadrons led gradually to the Cabibbo Theory, introduced 20 years ago. A detailed theory of weak interactions other than the phenomenological V-A Theory such as the Weinberg-Salam one was brought into very promising shape 10 years ago[2]. Clearly, progress in this area of Physics must be measured in decades and not just in years.

Despite the fact that β decay is a slowly moving field, it has led to very important discoveries, both theoretical and experimental ones. The apparent non-conservation of energy and angular momentum in nuclear β decay led to the neutrino. Today, the neutrino is probably the most delicate probe available to study the structure of matter and it may be a key particle to understand the gravitational stability of galaxies. The relevance of currents in high evergy physics, beyond Quantum Electrodynamics, was motivated by β decay. The technical difficulties of the non-renormalizability of the V-A Theory led to the advent of gauge theories in high energy physics and to, at least, a partial unification of two of the fundamental interactions in nature —weak and electromagnetic ones. β decay has played a relevant role as a source of new ideas and discoveries.

Nevertheless, many questions remain to be answered. Just to mention two of them, today, more than 20 years after the establishing of SU(3) symmetry, we still do not have a theory of symmetry breaking and also the Cabibbo angle has remained unrelated to other physical parameters. Hardly can one expect the latter to be a fundamental constant of nature. Important questions remain and β decay may help to solve them.

In a loose manner of speaking, we can say that β decay operates as a microscope by allowing the electron-neutrino pair to probe matter. That is, weak interactions when an electron and neutrino are present provide us with a tool, other than quantum electrodynamics, to explore in detail small regions where hadronic matter is present. This should be contrasted with weak interactions when only hadrons are present. It is because of this that β decay is bound to keep playing a relevant role.

The present state-of-the-art allows high statistics hyperon semileptonic experiments to be performed, experiments with hundreds of thousands and even millions of events. Such experiments would determine very precisely the strong-interaction form factors that appear in the corresponding matrix elements. The information thus obtained would be most helpful in guiding the theoretical work towards a theory of strong interactions. But, in order to fully exploit hyperon semileptonic decays, we must first see to what extent we understand them, and second we must learn how to make better use of them.

Our first aim in these lecture notes is to analyze in great detail what is known today on hyperon semileptonic decays. We shall pay much attention not to waste the rather limited available experimental information and not to make comparisons between theory and experiment that can be misleading because of simplifying approximations which are traditionally made, but which otherwise can be properly corrected. Our second aim is to set up a framework that is general and sound enough so that it can be used in the experiments to be performed in the next two decades.

The order of the material will reflect both aims simultaneously. This will allow us to introduce our framework as we use it. Only at the end shall we separate them. In the first part of these notes, Chapters 2, 3, and 4, we seek to answer the question whether the traditionally accepted description of hyperon semilptonic decays —namely, the Cabibbo theory— is in agreement or not with experiment. In the second part, Chapters 5, 6, and 7, we look for genuine discrepancies between theory and experiment, those that can not be attributed to the simplifying assumptions (or working assumptions as we shall call them later). In the third part, Chapters 8, and 9, we shall analyze in detail the meaning of the remaining discrepancies. As we shall see, two strong deviations will appear with very recently published data. One can be explained within the Cabibbo theory by small corrections due to symmetry breaking or to a non-pure, but still quite pure, octet axial-vector current. The other one can not be explained at all. Although at first sight this might seem to represent a "devastating" contradiction to the Cabibbo Theory, we shall see that it may not necessarily be so. What it may imply is that the use of the symmetry limit in a world where internal symmetries are broken must be revised. An approach that defines the symmetry limit in accordance with the basic postulates of Quantum Mechanics, from the outset, may be what this discrepancy with Cabibbo theory means. This is the approach of the Spectrum Generating models, which we study in Chapter 9. The Cabibbo theory is perfectly compatible with such an approach and the essential spirit of the latter theory can be preserved even in front of what seems to be a major contradiction. In the last, fourth part, limited to Chapter 10, we finally complete our frame-work and show how it can be used in high statistics experiments whose goals are measuring form factors to 1% or less.

Our conventions will be clarified throughout the text. We have numbered equations, tables, figures and references in simple order of appearance chapter by chapter. But when reference is made to an equation, table, etc. of another chapter, then the corresponding number takes the number of the chapter in front.

Chapter 2. Hyperon Semileptonic Decays

2.1. Features of the Decays.

There are many important differences between hyperon semileptonic decays (HSD) and other β decay processes. The fact that hyperons carry spin 1/2 makes the hadronic part of their semileptonic-decay matrix-elements richer in information on strong interactions than the corresponding ones of spin-zero meson leptonic and semileptonic decays. The mass difference between the hyperons requires that more form factors be taken into account than in nuclear or in free neutron β decay. In addition, hyperons carry strangeness —which the nucleons do not— and in many cases they carry different isopin assignments than the nucleons. Also, the emission of a μ-ν_μ pair instead of an e-ν_e pair is allowed in many HSD. It is clear then that HSD are more interesting processes than meson and nucleon β-decays.

There is, however, a very important limitation. Whereas β decay is the only decay mode of the neutron and it represents a very substantial fraction of meson decay channels, for hyperons HSD represent only a fraction of a percent of their several decay modes. This makes HSD difficult processes to observe.

We shall concentrate ourselves on the HSD that occur in the hyperon octet of SU(3). But our approach and our main formulas will be applicable to spin 1/2 hyperons that carry new quantum numbers such as c, b, or, the as yet undetected, top or t. In Table 1, we have listed all the β decays allowed by energy-momentum conservation within the baryon octet. Fig. 1 gives an schematic representation of the decays of Table 1. One can see that there are 16 allowed processes with the emission of an e-ν pair and 10 more with an emitted μ-ν pair —the appropiate electron or positron, anti-neutrino or neutrino should be understood, and similarly the appropiate μ^-, μ^+, ν_μ or $\bar{\nu}_\mu$ for μ-modes. In order to appreciate the energy release involved in each decay, we have also listed the mass difference ΔM involved and the parameter $\beta \equiv \Delta M/M_1$ (M_1 is the mass of the decaying hyperon). It is β and not ΔM that gives a better idea of how important recoil effects may be. It has a wide range of variation, over two orders of magnitude from $n \to pe\nu$ to $\Sigma^- \to ne\nu$ and $\Xi^- \to ne\nu$. Thus in $n \to pe\nu$ recoil effects are tiny, while in $\Sigma^- \to ne\nu$ they are important and should not be ignored. In the fourth and fifth columns of Table 1, we have also listed the changes in strangeness ΔS and charge ΔQ between the final baryon and the initial one. The relative sign and magnitude of these changes are related to selection rules that seem to operate in HSD. We shall return to this point in Chapter 4.

There are clearly many HSD within the baryon octet, 26 in all. We shall now turn to the general weak interaction theory that is believed to govern these decays.

Process	ΔM	β	ΔS	ΔQ	μ-mode
1. $n \to pe\nu$	1.29	0.0014	0	1	no
2. $\Sigma^+ \to \Lambda e\nu$	73.80	0.0620	0	-1	no
3. $\Sigma^- \to \Lambda e\nu$	81.70	0.0680	0	1	no
4. $\Lambda \to pe\nu$	177.30	0.1590	1	1	yes
5. $\Sigma^- \to ne\nu$	257.80	0.2150	1	1	yes
6. $\Xi^- \to \Lambda e\nu$	205.70	0.1560	1	1	yes
7. $\Xi^- \to \Sigma^0 e\nu$	128.90	0.0980	1	1	yes
8. $\Sigma^0 \to pe\nu$	254.20	0.2130	1	1	yes
9. $\Xi^0 \to \Sigma^+ e\nu$	125.50	0.0960	1	1	yes
10. $\Xi^- \to \Xi^0 e\nu$	6.42	0.0049	0	1	no
11. $\Sigma^- \to \Sigma^0 e\nu$	4.88	0.0041	0	1	no
12. $\Sigma^0 \to \Sigma^+ e\nu$	3.10	0.0026	0	1	no
13. $\Xi^- \to ne\nu$	381.80	0.2890	2	1	yes
14. $\Xi^0 \to pe\nu$	376.60	0.2860	2	1	yes
15. $\Sigma^+ \to ne\nu$	249.80	0.2100	1	-1	yes
16. $\Xi^0 \to \Sigma^- e\nu$	117.60	0.0890	1	-1	yes

TABLE 1. Processes energetically allowed within the hyperon octet. ΔM is the mass difference of the hyperons involved in the corresponding decay. ΔS is the strangeness change, final minus initial S. ΔQ is the charge change, again final minus initial Q. The last column indicates if the emission of a μ-ν_μ pair is allowed or not. No distinction has been made for the charge of the electron or positron emitted, nor for the kind of accompanying neutrino.

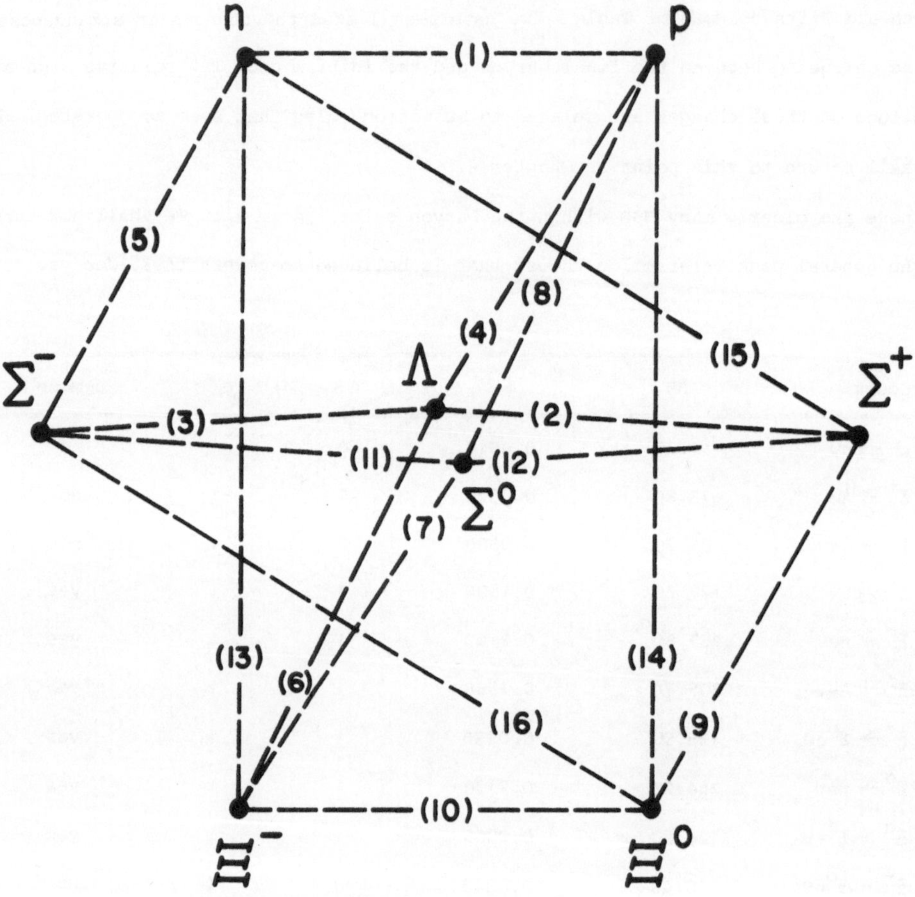

FIG. 1. Schematic diagram of the energetically allowed semileptonic decays within the hyperon octet.

2.2. Effective V-A Theory.

The original current-current interaction introduced by Fermi to describe nuclear β decay, which was later extended to incorporate parity violation[1], faces many problems. It does not have a renormalizable perturbation expansion and, thus, cannot produce finite higher order corrections with a finite number of counterterms as quantum electrodynamics does. At high energies it leads to a violation of unitarity; i.e., it brings in probability non-conservation. Being a point-like interaction, this so called V-A Theory[2] does not say anything about the possible existence of an intermediate boson[3], the analogous to the photon and the pion in electromagnetic and strong interactions, respectively. Yukawa himself suggested that such an intermediate vector boson should exist. The V-A Theory as was introduced by Sudarshan and Marshak and by Feynman and Gell-Mann applies only to processes where ΔQ is non-zero. It gives no indication as to how to extend it to describe processes with $\Delta Q = 0$. In other words the vector and axial-vector currents, V_μ and A_μ, are always charged. There is no room for neutral currents. Limited as it may be, the V-A theory has met enough success to make it what seems to be a sound starting point. It is at least consistent with a vast amount of experimental information on many nuclear β decays and with free neutron decay. Also, and this is probably its most attractive feature, it shows that weak interactions involving an e-ν or an μ-ν pair are new kinds of microscopes that can explore very small regions containing hadronic matter. There is no question about the usefulness of the V-A Theory both in describing many experimental data and in guiding and challenging physicists to look for a genuine theory of weak interactions.

With the advent of gauge theories[4], we can claim now that the ground has already been broken to finally establish the foundations of such a genuine theory of weak interactions. Today, everything points out that the $SU(2)\times U(1)$ Weinberg-Salam model constitutes the theory that describes the elementary processes of weak interaction transitions. We know at least that the Weinberg-Salam model provides an example of a field theory with a renormalizable perturbation expansion, with no unitarity problems, and with acceptable high energy behavior. This model operates between elementary particles, i.e., structureless particles; namely leptons, neutral and charged ones, and the fractionally charged quarks that are believed to exist and to be the elementary constituents of hadrons. Not only a charged intermediate vector boson $W^{\pm}$ is contained in the Weinberg-Salam model, but two neutral vector bosons, a massive one Z_o and a massless one γ. When the latter is identified with the photon the weak interactions are unified to a good extent with electromagnetic interactions. We say to a good extent because the $SU(2) \times U(1)$ model, containing a direct product, requires the existence of two coupling constants. Only one is determined using the electric charge e and the Fermi coupling constant G_μ, the other one is determined from experiment. One of them can be put in terms of an angle, the Weinberg angle θ_w.

Accepting this model as the elementary theory of weak interactions, the interaction hamiltonian reduces for low energy weak processes to an effective current-current interaction hamiltonian,

$$H_{int} = \frac{4G_\mu}{\sqrt{2}} \left(j_\mu j^\mu + 2J_\mu J^\mu\right), \tag{1}$$

where j_μ stands for the charged current and J_μ for the neutral current. If the two dimensionless coupling constants in this model are g and g', the electric charge is given by

$$e = g \sin \theta_w \tag{2}$$

and the Fermi constant by

$$G_\mu = \sqrt{2}\, g^2/8M_w^2 . \tag{3}$$

The masses of the charged and neutral bosons are related by

$$M_w = M_z \cos \theta_w . \tag{4}$$

The second coupling g' is replaced by θ_w through the relationship

$$\tan \theta_w = g'/g \tag{5}$$

Using the experimental values of[1.2]

$$G_\mu = (1.16632 \pm 0.00004) \times 10^{-5}\ \mathrm{GeV}^{-2} \qquad (6)$$
$$= 1.43582 \times 10^{-49}\ \mathrm{erg.\ cm}^3,$$

$$\alpha \equiv e^2/4\pi = 1/137.03604, \qquad (7)$$

and

$$\sin^2\theta_W \simeq 0.23, \qquad (8)$$

the masses of $W^\pm$ and Z_o are[1.2]

$$M_W \simeq 82\ \ \mathrm{GeV}, \qquad (9)$$

and

$$M_Z \simeq 91.5\ \ \mathrm{GeV} \qquad (10)$$

It is interesting to notice that although the above interaction hamiltonian is effectively reduced to the V-A Theory for $\Delta Q \neq 0$ transitions, it is no longer pure V-A for $\Delta Q = 0$ processes.

The current experimental evidence in favor of the SU(2) × U(1) model is quite impressive. Not only have neutral currents been discovered, but $W^\pm$ and Z_o have been discovered as well, with the masses very close to the values (9) and (10). The final test would be the coining down of the Higgs particle, which is brought into the game by the spontaneous breaking of the weak isospin symmetry. In this model this particle is supposed to be an isodoublet. Its discovery will require still a good deal of both experimental and theoretical effort. It is possible that there are several Higgs particles with a different weak isospin assignment or that more than one neutral intermediate boson exists. In this case the original Weinberg-Salam model should be replaced by a more elaborate one. Let us stress that it would not be fair to test the SU(2) × (1) model by the existence or non-existence of quarks and their properties. The problem of quarks belongs to the realm of strong interaction.

Attractive as it is, the SU(2) × U(1) model cannot be used directly in HSD. These decays are really the combination of three interactions; namely, weak, electromagnetic and strong interactions. The weak hamiltonian in Eq. (1) applies to the constituents of hadrons, but in HSD it is the hadrons themselves that show up. At

low energies the quark structure is hardly seen. How the quarks bind together to form
a hadron is so far an unsolved question. Non-perturbative strong interaction effects
at low energy weak interactions force one to use phenomenological form factors
to account for strong interaction effects in the matrix elements. The quarks
only play a qualitative role in HSD, justifying selection rules and giving hints
for the assignment of internal symmetry properties of the hadronic weak currents.
We are thus led back into an effective V-A Theory. Nevertheless, we shall find
the SU(2) × U(1) model useful when we discuss the electromagnetic corrections
to HSD. There it can provide some estimates of more than just qualitative value.
It is this effective V-A Theory that amounts to a microscope. The lepton pair
allows to determine the form factors that appear in the hadronic matrix element.
Their determination can furnish important guidance to determine how the quarks
bind together to form hadrons.

For HSD the transition amplitude in this effective V-A Theory is given by

$$M_0 = \frac{G_v}{\sqrt{2}} W_\lambda(p_1, p_2) \bar{u}_e \gamma_\lambda (1 + \gamma_5) v_\nu , \tag{11}$$

where

$$W_\lambda(p_1, p_2) = \bar{u}_B \left\{ f_1(q^2)\gamma_\lambda + \frac{f_2(q^2)}{M_1} \sigma_{\lambda\rho} q_\rho + \frac{f_3(q^2)}{M_1} q_\lambda + \right.$$
$$\left. + \left[g_1(q^2)\gamma_\lambda + \frac{g_2(q^2)}{M_1} \sigma_{\lambda\rho}q_\rho + \frac{g_3(q^2)}{M_1} q_\lambda \right] \gamma_5 \right\} u_A . \tag{12}$$

A and B refer to the decaying and emitted hyperons, respectively. The lepton part has
been written explicitly for e^- and $\bar{\nu}_e$, but the appropiate spinors for e^+ and ν_e ,
μ^- and $\bar{\nu}_\mu$ or μ^+ and ν_μ should be understood whenever necessary. Our metric and
γ matrices conventions are as in Ref. (5), except that our γ_5 has opposite sign and our
$\sigma_{\lambda\rho}$ does not carry an i. p_1, p_2, ℓ and p_ν are the momentum four vectors of A, B, e
and ν, respectively. Also, their masses will be denoted by M_1, M_2, m and zero. The
four-momentum transfer is defined as $q = p_1-p_2$. G_v is the vector-current coupling
constant, whose value in nuclear β decay is very close to G_μ of μ-decay.

In order for the induced form factors to have the same dimensions as the leading ones, we have divided them by M_1. f_2 is often referred to as the weak magnetism form factor and g_2 as the weak electric or pseudo-tensor form factor. f_3 and g_3 are the scalar and pseudo-scalar form factors. For electron or positron emission, these last two form factors will pick up a factor m in front of them as can be seen by moving the q_λ in front of them in Eq. (12) into the lepton covariant in Eq. (11). Using $q = \ell + p_\nu$ and the Dirac equation, their contribution to the transition rate will always be proportional to m and, therefore, will be negligible due to the smallness of this mass. For practical purposes then, the e-modes of HSD are described in terms of four, and not six, form factors; namely f_1, g_1, f_2, and g_2. For μ-modes, the ratio m/M_1 is still small, around 0.10, but may not be negligibly small and thus f_3 and g_3 may contribute with some significance.

f_2 and g_2 have very different properties if the V_μ and A_μ currents transform[6] in a definite way under charge conjugation $\mathcal{C}$ and charge symmetry (rotations in isospin space by $e^{i\pi I_2}$). Following Weinberg in putting these two operations together to form the G-parity, defined as

$$G = \mathcal{C} e^{i\pi I_2}, \tag{13}$$

we shall classify the weak currents as first-class currents if

$$\begin{aligned} G\, V_\mu G^{-1} &= V_\mu\,, \\ G\, A_\mu G^{-1} &= -A_\mu\,, \end{aligned} \tag{14}$$

and as second-class currents if

$$\begin{aligned} G\, V_\mu G^{-1} &= -V_\mu\,, \\ G\, A_\mu G^{-1} &= A_\mu\,. \end{aligned} \tag{15}$$

Then, for a HSD within two members of the same isomultiplet and in the limit in which time reversal and isospin symmetry are exact, first-class currents require for the induced form factors $g_2 = 0$, $f_3 = 0$, while second-class currents imply $f_2 = 0$, $g_3 = 0$. Because of this g_2 and f_3 are very often referred to as second-class form factors, and f_2 and g_3, as a first class ones. This is of course an informal manner of speaking. The first and second class properties, if they exist at all, are operator properties and not transition amplitude properties. For example, A_μ may be first class exclusively and still g_2 may be non-zero. As a matter of fact, in as much as isospin and SU(3) symmetries are broken symmetries it is expected that g_2 and f_3 be non-zero for pure first-class currents.

Although we must renounce to the use of the very simple and attractive SU(2)×U(1) model, we can use the effective V-A Theory with some confidence, knowing that, as long as strong interactions are well behaved, a field theory of quarks and leptons lies behind it. The real attraction for using the V-A Theory is that it allows Lorentz covariance and other generally accepted transformation properties to be fully exploited and one can thus keep full track of the effects of strong interactions in HSD. The existence of Z_o and $W^\pm$, i.e., the details of the elementary weak interaction theory, would amount to tiny contributions to the form factors, through higher order weak interaction corrections. As we shall see, it will be only through the radiative corrections to Eq. (11) that $W^\pm$ and Z_o may have a noticeable effect. Let us now turn to the transition probabilities for HSD.

2.3. Differential Decay Rate.

Long and tedious but otherwise straight forward calculations[7] starting with M_o in Eq. (11) give the differential decay rate of HSD. We shall give it for two cases; first when the initial hyperon is polarized along the direction s_1, and second when the polarization of the emitted hyperon is observed parallel to s_2. In the first case, the transition probability per unit time in the center of mass of A and with the charged lepton ℓ and neutrino going into the solid angles $d\Omega_\ell$ and $d\Omega_\nu$, respectively, is

$$d\omega(A\rightarrow B\ell\nu) = \frac{G_V^2}{(2\pi)^5} \frac{E\ell(E_m-E)^2 dE d\Omega_\ell d\Omega_\nu}{\left(1-\frac{E}{M_1}+\frac{\ell}{M_1}\hat{\ell}\cdot\hat{p}_\nu\right)^3} \left\{D_1 + \beta\hat{\ell}\cdot\hat{p}_\nu D_2 + \beta\hat{s}_1\cdot\hat{\ell}D_3 + \hat{s}_1\cdot\hat{p}_\nu D_4\right\} \qquad (16)$$

E_m is the charged lepton maximum energy; explicitly, $E_m = (M_1^2 - M_2^2 + m^2)/2M_1$. E and ℓ are its energy and three-momentum magnitude. β is here $\beta = \ell/E$ and it should not be confused with the β defined in Sec. 2.1. $\hat{\ell}$ and $\hat{p}_\nu$ are the directions of emissions of ℓ and ν, respectively. D_1, D_2, D_3, and D_4 are quadratic functions of the different form factors. They are explicitly

$$
\begin{aligned}
D_1 = {} & \left(2 - \frac{M_2}{M_1} - \frac{E}{M_1} - \frac{E_\nu}{M_1} - \frac{m^2}{M_1 E}\right) |F_1|^2 + \frac{1}{2}\left(1 + \frac{M_2}{M_1} - \frac{E}{M_1} - \frac{E_\nu}{M_1}\right) |F_2|^2 \\
& + \left(1 + \frac{M_2}{M_1} - \frac{m^2}{M_1 E}\right) \mathrm{Re}\, F_1 F_2^* + \frac{m^2}{M_1 E}\left(1 - \frac{E}{M_1} + \frac{M_2}{M_1}\right) \mathrm{Re}\, F_1 F_3^* + \\
& + \frac{m^2}{M_1 E}\left(1 + \frac{M_2}{M_1} - \frac{E_\nu}{M_1} - \frac{E}{M_1}\right) \mathrm{Re}\, F_2 F_3^* + \frac{1}{2}\frac{m^2}{M_1^2}\left(1 + \frac{M_2}{M_1} - \frac{E}{M_1} - \frac{E_\nu}{M_1}\right) |F_3|^2 + \\
& + \left(2 + \frac{M_2}{M_1} - \frac{E}{M_1} - \frac{E_\nu}{M_1} - \frac{m^2}{M_1 E}\right) |G_1|^2 + \frac{1}{2}\left(1 - \frac{M_2}{M_1} - \frac{E}{M_1} - \frac{E_\nu}{M_1}\right) |G_2|^2 + \\
& + \left(-1 + \frac{M_2}{M_1} + \frac{m^2}{M_1 E}\right) \mathrm{Re}\, G_1 G_2^* - \frac{m^2}{M_1 E}\left(1 - \frac{M_2}{M_1} - \frac{E}{M_1}\right) \mathrm{Re}\, G_1 G_3^* + \\
& + \frac{m^2}{M_1 E}\left(1 - \frac{M_2}{M_1} - \frac{E}{M_1} - \frac{E_\nu}{M_1}\right) \mathrm{Re}\, G_2 G_3^* + \frac{1}{2}\frac{m^2}{M_1^2}\left(1 - \frac{M_2}{M_1} - \frac{E}{M_1} - \frac{E_\nu}{M_1}\right) |G_3|^2 + \\
& + 2\left(\frac{E}{M_1} - \frac{E_\nu}{M_1} - \frac{m^2}{M_1 E}\right) \mathrm{Re}\, F_1 G_1^* ,
\end{aligned}
\tag{17}
$$

$$
\begin{aligned}
D_2 = {} & \left(\frac{E_\nu}{M_1} + \frac{E}{M_1} + \frac{M_2}{M_1}\right) |F_1|^2 + \frac{1}{2}\left(1 + \frac{M_2}{M_1} - \frac{E}{M_1} - \frac{E_\nu}{M_1}\right) |F_2|^2 + \\
& + \left(1 + \frac{M_2}{M_1}\right) \mathrm{Re}\, F_1 F_2^* + \frac{m^2}{M_1^2} \mathrm{Re}\, F_1 F_3^* + \frac{1}{2}\frac{m^2}{M_1^2}\left(-1 - \frac{M_2}{M_1} + \frac{E}{M_1} + \frac{E_\nu}{M_1}\right) |F_3|^2 + \\
& + \left(-\frac{M_2}{M_1} + \frac{E}{M_1} + \frac{E_\nu}{M_1}\right) |G_1|^2 + \frac{1}{2}\left(1 - \frac{M_2}{M_1} - \frac{E}{M_1} - \frac{E_\nu}{M_1}\right) |G_2|^2 + \\
& + \left(-1 + \frac{M_2}{M_1}\right) \mathrm{Re}\, G_1 G_2^* - \frac{m^2}{M_1^2} \mathrm{Re}\, G_1 G_3^* + \\
& + \frac{1}{2}\frac{m^2}{M_1^2}\left(-1 + \frac{M_2}{M_1} + \frac{E}{M_1} + \frac{E_\nu}{M_1}\right) |G_3|^2 + 2\left(-\frac{E}{M_1} + \frac{E_\nu}{M_1}\right) \mathrm{Re}\, F_1 G_1^* ,
\end{aligned}
\tag{18}
$$

$$D_3 = \left(-1 + \frac{M_2}{M_1} + \frac{E}{M_1} - X_\ell\right) |F_1|^2 + \left(-\frac{E_\nu}{M_1} - X_\ell\right) \mathrm{Re}\, F_1F_2^* +$$

$$+ \left(-1 - \frac{M_2}{M_1} + \frac{E}{M_1} - X_\ell\right) |G_1|^2 + \left(\frac{E_\nu}{M_1} + X_\ell\right) \mathrm{Re}\, G_1G_2^* +$$

$$+ 2\left(1 - \frac{E}{M_1} + X_\ell\right) \mathrm{Re}\, F_1G_1^* + \left(-1 + \frac{M_2}{M_1} + \frac{E_\nu}{M_1} - X_\ell\right) \mathrm{Re}\, F_1G_2^* +$$

$$+ \left(1 + \frac{M_2}{M_1} - \frac{E_\nu}{M_1} + X_\ell\right) \mathrm{Re}\, G_1F_2^* - \frac{m^2}{M_1^2} \mathrm{Re}\, F_1G_3^* + \frac{m^2}{M_1^2} \mathrm{Re}\, G_1F_3^* -$$

$$- \left(\frac{E}{M_1} + X_\ell\right) \mathrm{Re}\, F_2G_2^* - \frac{m^2}{M_1^2} \mathrm{Re}\, F_2G_3^* - \frac{m^2}{M_1^2} \mathrm{Re}\, F_3G_2^* +$$

$$+ \frac{m^2}{M_1^2} \left(-\frac{E}{M_1} + X_\ell\right) \mathrm{Re}\, F_3G_3^* , \qquad (19)$$

$$D_4 = \left(1 - \frac{M_2}{M_1} - \frac{E_\nu}{M_1} - \frac{m^2}{M_1E} + X_\nu\right) |F_1|^2 + \left(\frac{E}{M_1} - \frac{m^2}{M_1E} + X_\nu\right) \mathrm{Re}\, F_1F_2^* +$$

$$+ \left(1 + \frac{M_2}{M_1} - \frac{E_\nu}{M_1} - \frac{m^2}{M_1E} + X_\nu\right) |G_1|^2 + \left(-\frac{E}{M_1} + \frac{m^2}{M_1E} - X_\nu\right) \mathrm{Re}\, G_1G_2^* +$$

$$+ 2\left(1 - \frac{E_\nu}{M_1} - \frac{m^2}{M_1E} + X_\nu\right) \mathrm{Re}\, F_1G_1^* + \left(-1 + \frac{M_2}{M_1} + \frac{E}{M_1} - X_\nu\right) \mathrm{Re}\, F_1G_2^* +$$

$$+ \left(1 + \frac{M_2}{M_1} - \frac{E}{M_1} + X_\nu\right) \mathrm{Re}\, G_1F_2^* + \frac{m^2}{M_1E} \left(-1 + \frac{M_2}{M_1} + \frac{E}{M_1}\right) \mathrm{Re}\, F_1G_3^* +$$

$$+ \frac{m^2}{M_1E} \left(1 + \frac{M_2}{M_1} - \frac{E}{M_1}\right) \mathrm{Re}\, G_1F_3^* - \left(\frac{E_\nu}{M_1} + X_\nu\right) \mathrm{Re}\, F_2G_2^*$$

$$- \frac{m^2}{M_1E} \frac{E_\nu}{M_1} \mathrm{Re}\, F_2G_3^* - \frac{m^2}{M_1E} \frac{E_\nu}{M_1} \mathrm{Re}\, F_3G_2^* + \frac{m^2}{M_1^2} \left(-\frac{E_\nu}{M_1} + X_\nu\right) \mathrm{Re}\, F_3G_3^* . \qquad (20)$$

We used the definitions

$$X_\ell = \hat{\ell}\cdot\hat{p}_\nu \frac{\ell}{M_1} , \qquad (21)$$

$$X_\nu = \beta\hat{\ell}\cdot\hat{p}_\nu \frac{E_\nu}{M_1} , \tag{22}$$

$$F_1 = f_1 + \left(1 + \frac{M_2}{M_1}\right) f_2 , \quad G_1 = g_1 - \left(1 - \frac{M_2}{M_1}\right) g_2$$

$$F_2 = -2f_2 , \quad G_2 = -2g_2 , \quad F_3 = f_2 + f_3 , \quad G_3 = g_2 + g_3 \tag{23}$$

The neutrino energy is given by $E_\nu = (E_m - E)/(1 - \frac{E}{M_1} + X_\ell)$. Although the coefficients D_i $(i = 1,2,3,4)$ look breathtaking, still the decay rate Eq. (16) has a very simple form. It consists of only four terms with the phase-space coefficient in front of them.

When the polarization of the emitted hyperon is observed, the differential decay rate in the center of mass of the emitted hyperon B is

$$dw(A \to B\ell\nu) = \frac{1}{2}\left(\frac{M_2}{M_1}\right)^3 \frac{G_V^2}{(2\pi)^5} \frac{(\hat{E}_m - \hat{E})^2 \hat{\ell}\hat{E}d\hat{E}d\hat{\Omega}_\ell d\hat{\Omega}_\nu}{\left[1 + \frac{\hat{E}}{M_2}(1 - \beta\hat{x})\right]^3} \times \{\hat{D}_1 + \beta\hat{\ell}^*\cdot\hat{p}_\nu^* \hat{D}_2$$

$$+ \beta\hat{s}_2\cdot\hat{\ell}^* \hat{D}_5 + \hat{s}_2\cdot\hat{p}_\nu^* \hat{D}_6\} , \tag{24}$$

The "hat" over the variables is just a reminder of the frame we are working in. $\hat{\ell}^*$ and $\hat{p}_\nu^*$ are the emission directions of ℓ and $\bar{\nu}_\ell$, $\hat{x} = \hat{\ell}^*\cdot\hat{p}_\nu^*$, and the maximum energy of ℓ is now $\hat{E}_m = (M_1^2 - M_2^2 - m^2)/2M_2$. The neutrino energy is now $\hat{E}_\nu = (\hat{E}_m - \hat{E})/[1 + \hat{E}(1 - \beta\hat{x})/M_2]$ β again is the velocity of ℓ, although we do not put a hat on top of it, it must be understood that it is given in the center of mass of B. We shall not give $\hat{D}_1$ and $\hat{D}_2$ explicitly since we shall not need them. $\hat{D}_5$ and $\hat{D}_6$ are given by

$$\hat{D}_5 = |F_1|^2\left(1 - \frac{M_1}{M_2} + \frac{\hat{E}}{M_2} - \beta\frac{\hat{E}}{M_2}\hat{x}\right) + |G_1|^2\left(1 + \frac{M_1}{M_2} + \frac{\hat{E}}{M_2} - \beta\frac{\hat{E}}{M_2}\hat{x}\right) +$$

$$+ 2\mathrm{Re}\, F_1G_1^*\left(1 + \frac{\hat{E}}{M_2} - \beta\frac{\hat{E}}{M_2}\hat{x}\right) + (\mathrm{Re}\, F_1F_2^* + \mathrm{Re}\, G_1G_2^*)\left(-\frac{\hat{E}_\nu}{M_1} - \beta\frac{\hat{E}}{M_1}\hat{x}\right) +$$

$$+ \text{Re}\ F_1G_2^*\left(-1 + \frac{M_2}{M_1} + \frac{\hat{E}_\nu}{M_1} - \beta\,\frac{\hat{E}}{M_1}\,\hat{x} - \frac{m^2}{M_1M_2}\right) +$$

$$+ \text{Re}\ G_1F_2^*\left(1 + \frac{M_2}{M_1} + \frac{\hat{E}_\nu}{M_1} - \beta\,\frac{\hat{E}}{M_1}\,\hat{x} - \frac{m^2}{M_1M_2}\right) +$$

$$+ \text{Re}\ F_2G_2^*\,\frac{M_2}{M_1}\left(-\frac{\hat{E}}{M_1} - \beta\,\frac{\hat{E}}{M_1}\,\hat{x} - \frac{2m^2}{M_1M_2} - \frac{m^2\hat{E}}{M_1M_2^2} + \beta\,\frac{m^2\hat{E}}{M_1M_2^2}\,\hat{x}\right) +$$

$$+ \left(\text{Re}\ F_1G_3^* + \text{Re}\ G_1F_3^*\right)\left(-\frac{m^2}{M_1M_2}\right) + \left(\text{Re}\ F_2G_3^* + \text{Re}\ F_3G_2^*\right)\frac{m^2}{M_1^2}\left(-1 - \frac{\hat{E}}{M_2} + \beta\,\frac{\hat{E}}{M_2}\,\hat{x}\right) +$$

$$+ \text{Re}\ F_3G_3^*\,\frac{m^2}{M_1^2}\left(-\frac{\hat{E}}{M_2} + \beta\,\frac{\hat{E}}{M_2}\,\hat{x}\right) \tag{25}$$

and

$$\hat{D}_6 = |F_1|^2\left(-1 + \frac{M_1}{M_2} - \frac{\hat{E}_\nu}{M_2} + \beta\,\frac{\hat{E}_\nu}{M_2}\,\hat{x} - \frac{m^2}{M_2\hat{E}}\right) +$$

$$+ |G_1|^2\left(-1 - \frac{M_1}{M_2} - \frac{\hat{E}_\nu}{M_2} + \beta\,\frac{\hat{E}_\nu}{M_2}\,\hat{x} - \frac{m^2}{M_2\hat{E}}\right) +$$

$$+ 2\text{Re}\ F_1G_1^*\left(1 + \frac{\hat{E}_\nu}{M_2} - \beta\,\frac{\hat{E}_\nu}{M_2}\,\hat{x} + \frac{m^2}{M_2\hat{E}}\right) +$$

$$+ (\text{Re}\ F_1F_2^* + \text{Re}\ G_1G_2^*)\left(\frac{\hat{E}}{M_1} + \beta\,\frac{\hat{E}_\nu}{M_1}\,\hat{x} - \frac{m^2}{M_1\hat{E}}\right) +$$

$$+ \text{Re}\ F_1G_2^*\left[-1 + \frac{M_2}{M_1} + \frac{\hat{E}}{M_1} - \beta\,\frac{\hat{E}_\nu}{M_1}\,\hat{x} + \frac{m^2}{M_1\hat{E}}\left(1 - \frac{M_1}{M_2}\right) + \frac{m^2}{M_1M_2}\right] +$$

$$+ \text{Re}\ G_1F_2^*\left[1 + \frac{M_2}{M_1} + \frac{\hat{E}}{M_1} - \beta\,\frac{\hat{E}_\nu}{M_1}\,\hat{x} + \frac{m^2}{M_1\hat{E}}\left(1 + \frac{M_1}{M_2}\right) + \frac{m^2}{M_1M_2}\right] +$$

$$+ \text{Re}\ F_2G_2^*\left(\frac{M_2}{M_1}\right)\left(-\frac{\hat{E}_\nu}{M_1} - \beta\,\frac{\hat{E}_\nu}{M_1}\,\hat{x} - \frac{m^2\hat{E}_\nu}{M_2^2M_1} + \beta\,\frac{\hat{E}_\nu}{M_1}\,\frac{m^2}{M_2^2}\,\hat{x} - \frac{2\hat{E}_\nu m^2}{\hat{E}M_1M_2}\right) +$$

$$+ \text{Re } F_1 G_3^* \left(\frac{m^2}{M_1 \hat{E}}\right) \left(1 - \frac{M_1}{M_2} + \frac{\hat{E}}{M_2}\right) +$$

$$+ \text{Re } G_1 F_3^* \left(\frac{m^2}{M_1 \hat{E}}\right) \left(1 + \frac{M_1}{M_2} + \frac{\hat{E}}{M_2}\right) +$$

$$+ (\text{Re } F_2 G_3^* + \text{Re } F_3 G_2^*) \left(\frac{m^2}{M_1 \hat{E}}\right) \left(- \frac{\hat{E}_\nu}{M_1} - \frac{\hat{E}\hat{E}_\nu}{M_1 M_2} + \beta \frac{\hat{E}\hat{E}_\nu}{M_1 M_2} \hat{x}\right) +$$

$$+ \text{Re } F_3 G_3^* \left(\frac{m^2}{M_1^2}\right) \left(- \frac{\hat{E}_\nu}{M_2} + \beta \frac{\hat{E}_\nu}{M_2} \hat{x}\right) . \tag{26}$$

Again the length of D_5 and D_6 should not mask the basic simplicity of Eq. (24).

We have made no approximation as to the smallness of m in Eqs. (16) and (24). They apply then to the μ-modes also. We shall not need an expression for the case where both A and B are simultaneously polarized. It can be found in Ref. (8). Eq. (16) was calculated before by Harrington[9]; our result and his agree when m is neglected, but there are minor differences in some terms proportional to m. Eq. (24) was calculated before by Linke[8]; our result agrees completely with his.

We shall return to Eqs. (16) and (24) in Chapter 5, when we study the effect of radiative corrections. Let us now turn to the experimental evidence on HSD.

Chapter 3. Low Statistics Experiments

3.1. Integrated Observables.

As we mentioned in the last chapter, the main drawback of HSD is their small branching ratio, typically a fraction of a percent. Also, hyperons are not that abundant —they are usually produced in small quantities. These two limitations have made that the number of events of HSD be quite small; in the past only several hundreds of events have been collected for some HSD. This has led to low statistics experiments. Fortunately, this situation is gradually improving. Recently, several thousands have been already collected for certain decays and currently experiments with several tens of thousands are being analyzed. Even in free neutron decay experiments have low statistics. One should contrast this with meson leptonic and semileptonic decays, where collecting hundreds of thousands and even more than a million events has not been a problem for over a decade. The contrast is even more impressive if one remembers that in nuclear β decay collecting millions of events is a reasonable expectation.

Because of the low statistics one cannot perform a detailed analysis of the Dalitz plot, i.e., the differential decay rate Eq. (2.16). One must lump the events together to produce certain integrated observables; namely, the total transition rate R and angular correlation and asymmetry coefficients. The practice of forming these integrated observables can be traced back to the early times of the parity violation discovery[2.1]. It has remained the standard practice for presenting experimental results in free neutron decay. The advantage of these observables is that their definition does not assume any particular theoretical approach. They are defined using only kinematics; for example, the e-ν angular correlation coefficient is defined as

$$\alpha_{e\nu} = 2\,\frac{N\left(\theta_{e\nu} < \frac{1}{2}\pi\right) - N\left(\theta_{e\nu} > \frac{1}{2}\pi\right)}{N\left(\theta_{e\nu} < \frac{1}{2}\pi\right) + N\left(\theta_{e\nu} > \frac{1}{2}\pi\right)}\,, \tag{1}$$

where $N\left(\theta_{e\nu} \lessgtr \frac{1}{2}\pi\right)$ is the number of all events with e-ν pairs emitted in directions that make an angle between them smaller or greater than 90°, respectively. In a similar fashion the e-asymmetry coefficient is defined as

$$\alpha_e = 2\,\frac{N\left(\theta_e < \frac{1}{2}\pi\right) - N\left(\theta_e > \frac{1}{2}\pi\right)}{N\left(\theta_e < \frac{1}{2}\pi\right) + N\left(\theta_e > \frac{1}{2}\pi\right)}\,, \tag{2}$$

where θ_e is the angle between the e-direction and polarization direction of the decaying hyperon. Analogous definitions are used for the ν-asymmetry α_ν and the emitted hyperon asymmetry α_B. When the polarization of the recoil hyperon can be measured one can introduce new asymmetry coefficients, an e-asymmetry $\hat{\alpha}_e$ and a ν-asymmetry $\hat{\alpha}_\nu$. However, in this case, it has been proposed[1] to use somewhat different asymmetry coefficients by introducing an orthonormal basis (a' and b' are normalization factors),

$$\hat{\alpha} = (\hat{\ell}^* + \hat{p}_\nu^*)/a'\,, \tag{3}$$

$$\hat{\beta} = (\hat{\ell}^* - \hat{p}_\nu^*)/b'\,, \tag{4}$$

and

$$\hat{n} = \hat{\beta} \times \hat{\alpha}\,, \tag{5}$$

to replace the e- and ν- directions $\hat{\ell}^*$ and $\hat{p}_\nu^*$ in Eq. (2.24). Two asymmetry coefficients are defined using $\hat{\alpha}$ and $\hat{\beta}$ and s_2, in analogy with Eq. (2). The one related to $\hat{\alpha}$ we shall call A, and the one related to $\hat{\beta}$ we shall call B.

When the e-mass can be neglected one can produce approximate theoretical expressions for these angular coefficients rather easily[2]. For example, if the initial hyperon is polarized one has

$$R = G_V^2 \frac{\Delta m^5}{60\pi^3}\Big[\left(1 - \tfrac{3}{2}\beta + \tfrac{6}{7}\beta^2\right)|f_1|^2 + \left(\tfrac{4}{7}\beta^2\right)|f_2|^2 + \left(3 - \tfrac{9}{2}\beta + \tfrac{12}{7}\beta^2\right)|g_1|^2 + $$

$$+ \left(\tfrac{12}{7}\beta^2\right)|g_2|^2 + \left(\tfrac{6}{7}\beta^2\right)\mathrm{Re}\, f_1 f_2^* + (-4\beta + 6\beta^2)\,\mathrm{Re}\, g_1 g_2^*\Big]\,, \tag{6}$$

$$R \times \alpha_{e\nu} = G_V^2 \frac{\Delta m^5}{60\pi^3}\Big[\left(1 - \tfrac{5}{2}\beta + \tfrac{11}{7}\beta^2\right)|f_1|^2 + \left(-\tfrac{2}{7}\beta^2\right)|f_2|^2 + $$

$$+ \left(-1 - \tfrac{3}{2}\beta + \tfrac{25}{7}\beta^2\right)|g_1|^2 + (-2\beta^2)\,|g_2|^2$$

$$+ \left(-\frac{2}{7}\beta^2\right)\mathrm{Re}\, f_1f_2^* + (4\beta - 2\beta^2)\,\mathrm{Re}\, g_1g_2^*\big] \,. \tag{7}$$

$$R\times\alpha_e = G_V^2\,\frac{\Delta m^5}{60\pi^3}\,\Big[\left(-\frac{1}{3}\beta + \frac{3}{14}\beta^2\right)|f_1|^2 + \left(-\frac{4}{21}\beta^2\right)|f_2^2| + \left(-2 + \frac{8}{3}\beta - \frac{9}{14}\beta^2\right)|g_1|^2 +$$

$$+ \left(-\frac{4}{3}\beta^2\right)|g_2|^2 + \left(-\frac{2}{3}\beta + \frac{14}{21}\beta^2\right)\mathrm{Re}\, f_1f_2^* +$$

$$+ \left(2 - \frac{11}{3}\beta + \frac{15}{7}\beta^2\right)\mathrm{Re}\, f_1g_1^* + \left(-\frac{2}{3}\beta + \frac{32}{21}\beta^2\right)\mathrm{Re}\, f_1g_2^* +$$

$$+ \left(-\frac{2}{3}\beta + \frac{32}{21}\beta^2\right)\mathrm{Re}\, f_2g_1^* + \left(\frac{16}{21}\beta^2\right)\mathrm{Re}\, f_2g_2^* + \left(\frac{10}{3}\beta - \frac{94}{21}\beta^2\right)\mathrm{Re}\, g_1g_2^*\Big], \tag{8}$$

$$R\times\alpha_\nu = G_V^2\,\frac{\Delta m^5}{60\pi^3}\,\Big[\left(\frac{1}{3}\beta - \frac{3}{14}\beta^2\right)|f_1|^2 + \left(\frac{4}{21}\beta^2\right)|f_2^2| + \left(2 - \frac{8}{3}\beta + \frac{9}{14}\beta^2\right)|g_1|^2 +$$

$$+ \left(\frac{2}{3}\beta - \frac{14}{21}\beta^2\right)\mathrm{Re}\, f_1f_2^* + \left(2 - \frac{11}{3}\beta + \frac{15}{7}\beta^2\right)\mathrm{Re}\, f_1g_1^* + \left(\frac{4}{3}\beta^2\right)|g_2^2|$$

$$+ \left(-\frac{2}{3}\beta + \frac{32}{21}\beta^2\right)\mathrm{Re}\, f_1g_2^* + \left(-\frac{2}{3}\beta + \frac{32}{21}\beta^2\right)\mathrm{Re}\, f_2g_1^* +$$

$$+ \left(\frac{16}{21}\beta^2\right)\mathrm{Re}\, f_2g_2^* + \left(-\frac{10}{3}\beta + \frac{94}{21}\beta^2\right)\mathrm{Re}\, g_1g_2^*\Big] \tag{9}$$

$$R\times\alpha_B = G_V^2\,\frac{\Delta m^5}{60\pi^3}\,\frac{5}{2}\,\Big[\left(-1 + \frac{11}{6}\beta - \beta^2\right)\mathrm{Re}\, f_1g_1^* + \left(\frac{1}{3}\beta - \frac{5}{6}\beta^2\right)\mathrm{Re}\, f_1g_2^* +$$

$$+ \left(\frac{2}{3}\beta - \frac{7}{6}\beta^2\right)\mathrm{Re}\, f_2g_1^* + \left(-\frac{2}{3}\beta^2\right)\mathrm{Re}\, f_2g_2^*\Big], \tag{10}$$

β is the parameter of Table 2.1; namely, $\Delta M/M_1$. When β is very small, the above formulas take the very simple forms that have been in the literature[1.1] for a very long time. In Eqs. (6-10), the form factors have been assumed to be constant. Their q^2-dependence cannot always be neglected, since in many cases they give a noticeable contribution. But even when β is not very small, it is only $f_1(q^2)$ and $g_1(q^2)$ that count, while f_2, g_2, f_3 and g_3 can still be considered to be constant. A linear expansion in q^2 is enough, because higher powers of q^2 amount to contributions no larger than a fraction of a percent. The expansions of f_1 and g_1 are then

$$f_1(q^2) = f_1(0) + \frac{q^2}{M_1^2} \lambda_1^f \quad , \tag{11}$$

and

$$g_1(q^2) = g_1(0) + \frac{q^2}{M_1^2} \lambda_1^g \quad , \tag{12}$$

we have made the slope parameters λ_1^f and λ_1^g dimensionless by dividing q^2 by M_1^2. It is, of course expected, that λ_1^f and λ_1^g be numerically of order unity in accordance with analyticity properties of strong interactions. If this were not the case and they were too big, expansions (11) and (12) would make little sense. Eqs. (6-10) get the contributions (6'-10'), respectively:

$$R: \; G_V^2 \frac{\Delta m^5}{60\pi^3} \left(\frac{4}{7}\beta^2\right)\left(f_1\lambda_1^f + 5g_1\lambda_1^g\right), \tag{6'}$$

$$R\times\alpha_{e\nu}: \; G_V^2 \frac{\Delta m^5}{60\pi^3} \left(\frac{24}{7}\beta^2\right)\left(-g_1\lambda_1^g\right), \tag{7'}$$

$$R\times\alpha_e: \; G_V^2 \frac{\Delta m^5}{60\pi^3} \left(\frac{4}{7}\beta^2\right)\left(g_1\lambda_1^f + f_1\lambda_1^g - 4g_1\lambda_1^g\right), \tag{8'}$$

$$R\times\alpha_\nu: \; G_V^2 \frac{\Delta m^5}{60\pi^3} \left(\frac{4}{7}\beta^2\right)\left(g_1\lambda_1^f + f_1\lambda_1^g + 4g_1\lambda_1^g\right), \tag{9'}$$

$$R\times\alpha_B: \; G_V^2 \frac{\Delta m^5}{60\pi^3} \left(-\frac{5}{6}\beta^2\right)\left(f_1\lambda_1^g + g_1\lambda_1^f\right), \tag{10'}$$

where $f_1 = f_1(0)$ and $g_1 = g_1(0)$.

For more precise formulas[3] and, also, when m should be kept in, it is better to integrate numerically Eqs. (2.16) and (2.24). In this respect the usefulness of Eqs. (6-10) is to serve as a fast check on the numerical results. This is the main reason why we have reproduced them here. For comparisons between theory and experiment it is the numerical formulas we shall use. They take several pages, so we have placed them in the Appendix 1. There we also include the numerical formulas for A and B. For completeness, we have computed formulas for all the 16 electron mode and the 10 muon mode processes in Table 2.1.

Let us notice that R is a very sensitive function of ΔM. It is such a dependence that accounts for the typically nine orders of magnitude between the transition rate of the free neutron and the rates of the other hyperons. It is also interesting to see how some theorems of Weinberg[4] apply to Eqs. (6-10). Neither R nor $\alpha_{e\nu}$ contain interference terms with simultaneously vector and axial-vector form factors. Also, such interference terms appear in α_e and α_ν with the same sign, while others terms in these two observables appear with opposite signs, as they should. Let us remark that g_2 contributes to some of these expressions through terms Re $g_1 g_2^*$ with coefficients comparable to those of f_1^2. A similar remark had already been made[5] before for the electron energy spectrum and the decay rate. We shall appreciate this better in the next section.

To close this section, let us mention that another practice to present the experimental results is by quoting values for the g_1/f_1 ratios. This practice is, however, questionable. Looking through Eqs. (6-10) and through Appendix 1 it is clear that with only a few observables one cannot determine all the unknowns uniquely. Experimental values for these ratios can only be obtained in low statistics experiments if one neglects the other form factors or one fixes them by introducing theoretical predictions for them. Only for $n \to pe\nu$ is this practice satisfactory, because β is so small. But for other decays the values of g_1/f_1 are very model dependent.

3.2. Sensitivity to Form Factors.

It is difficult to see directly from Eqs. (6-10) how sensitive the angular coefficients are to the different form factor ratios because the effect of dividing these equations by R may be deceiving. It is better to plot them as functions of the different form factor ratios; namely, g_1/f_1, g_2/f_1, and f_2/f_1, when m is negligible. As an example, we have done[2,6] this in Figs. (1-6) for $\alpha_{e\nu}$ and α_e in $\Sigma^- \to ne\nu$. The

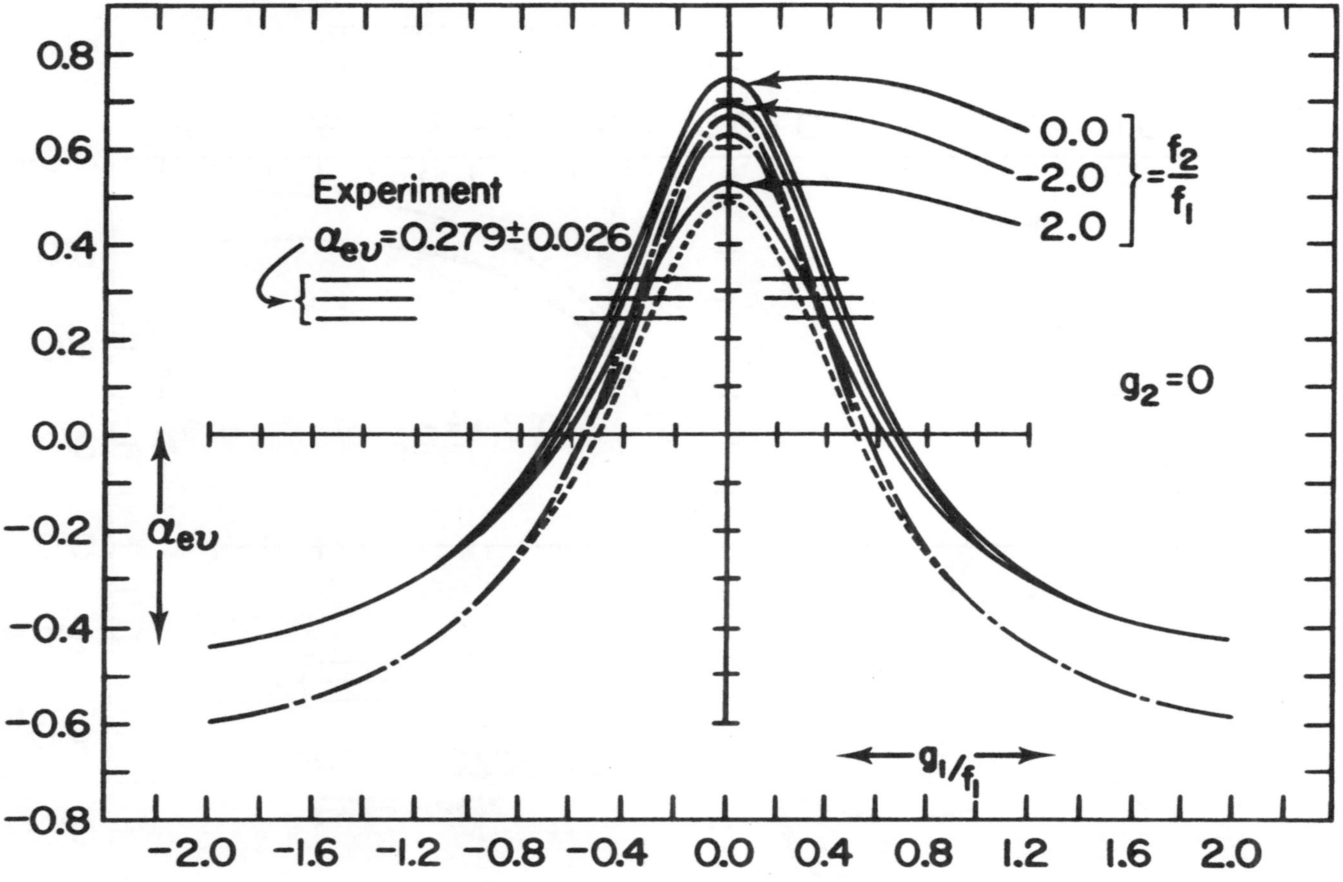

FIG. 1. Plots of $\alpha_{e\nu}$ in $\Sigma^- \rightarrow ne\nu$ as a function of g_1/f_1 and f_2/f_1. g_2 is kept fixed at zero. The dotted lines correspond to adding q^2-dependence of f_1 and g_1 at vector and axial-vector meson dominance.

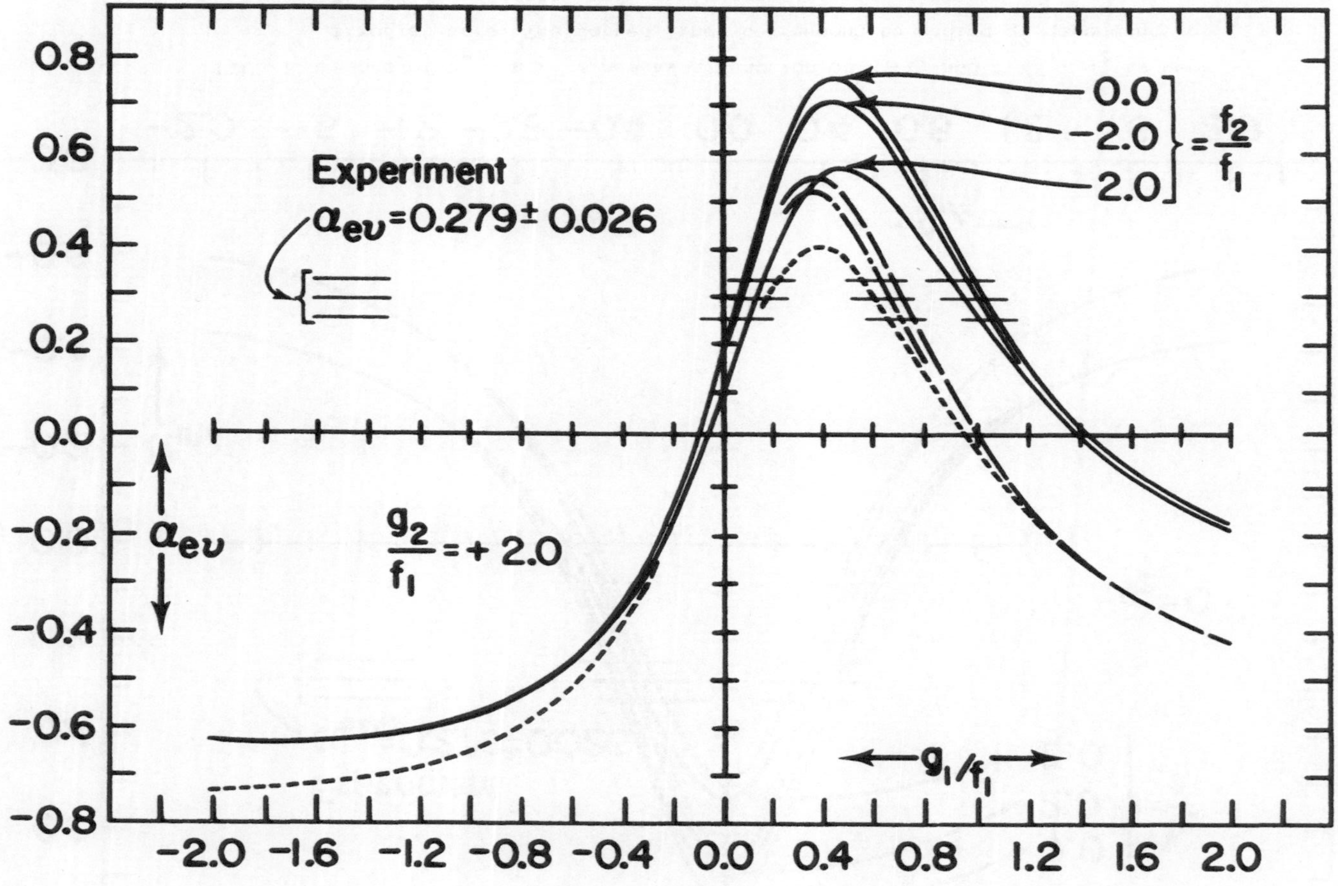

FIG. 2. Plots of $\alpha_{e\nu}$ in $\Sigma^- \rightarrow ne\nu$ as a function of g_1/f_1 and f_2/g_1. g_2 to kept fixed at 2.0. The dotted lines have the same meaning as in Fig. 1.

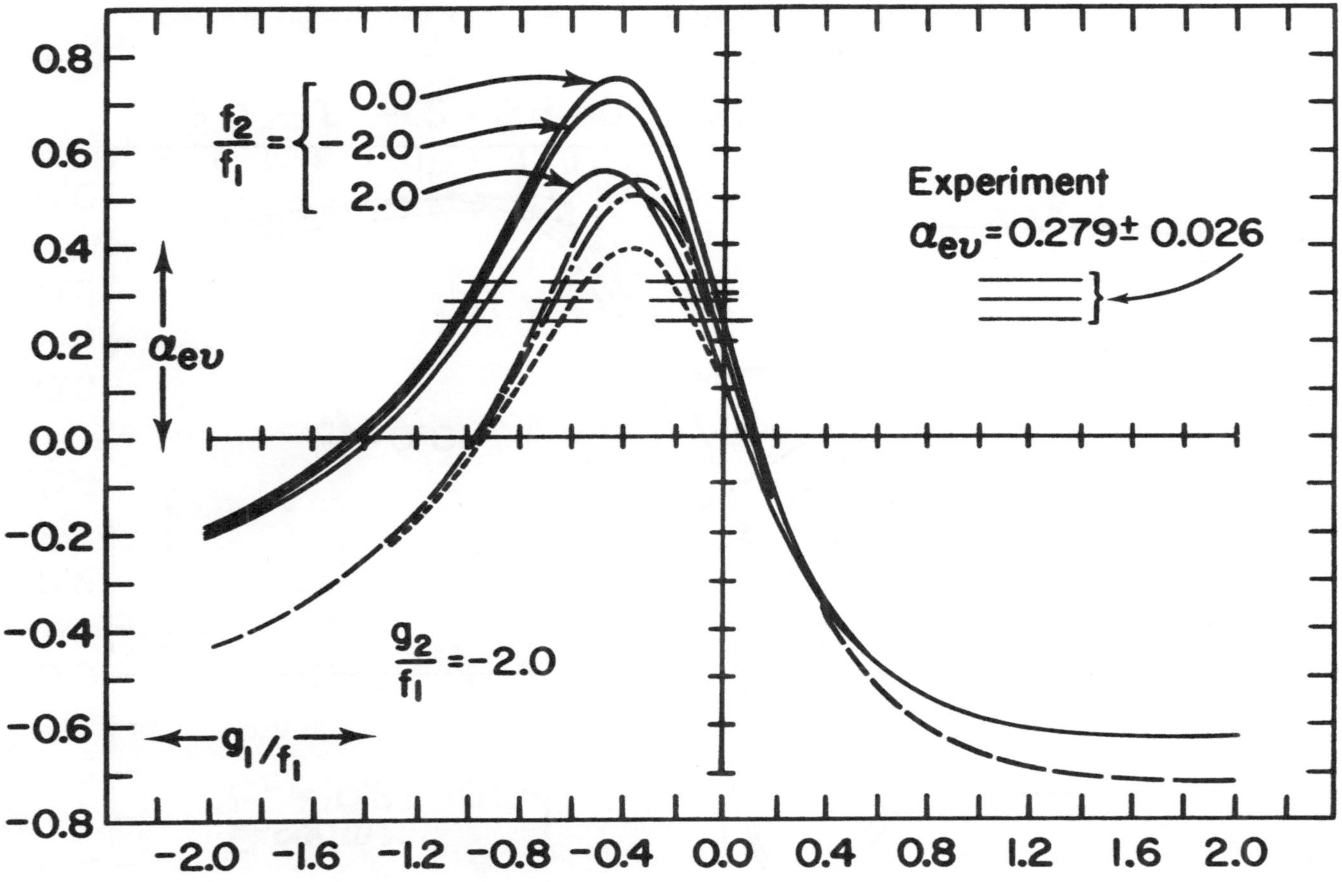

FIG. 3. Plots of $\alpha_{e\nu}$ of $\Sigma^- \to ne\nu$ as a function of g_1/f_1 and f_2/f_1. g_2 is kept fixed at -2.0. The dotted lines have the same meaning as in Fig. 1.

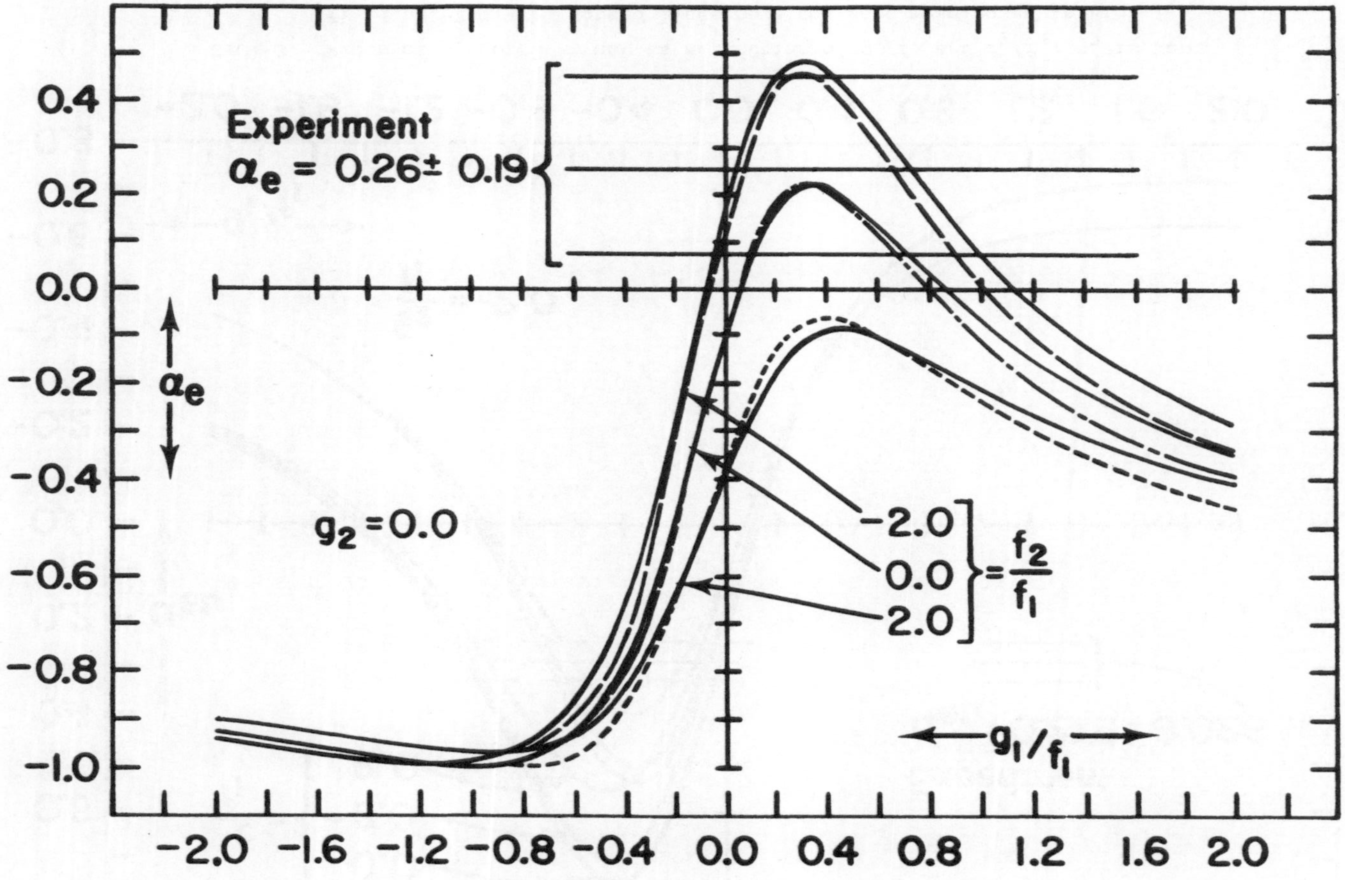

FIG. 4. Plots of α_e in $\Sigma^- \rightarrow ne\nu$ as a function of g_1/f_1 and f_2/f_1. g_2 is kept fixed at zero. The dotted lines have the same meaning as in Fig. 1.

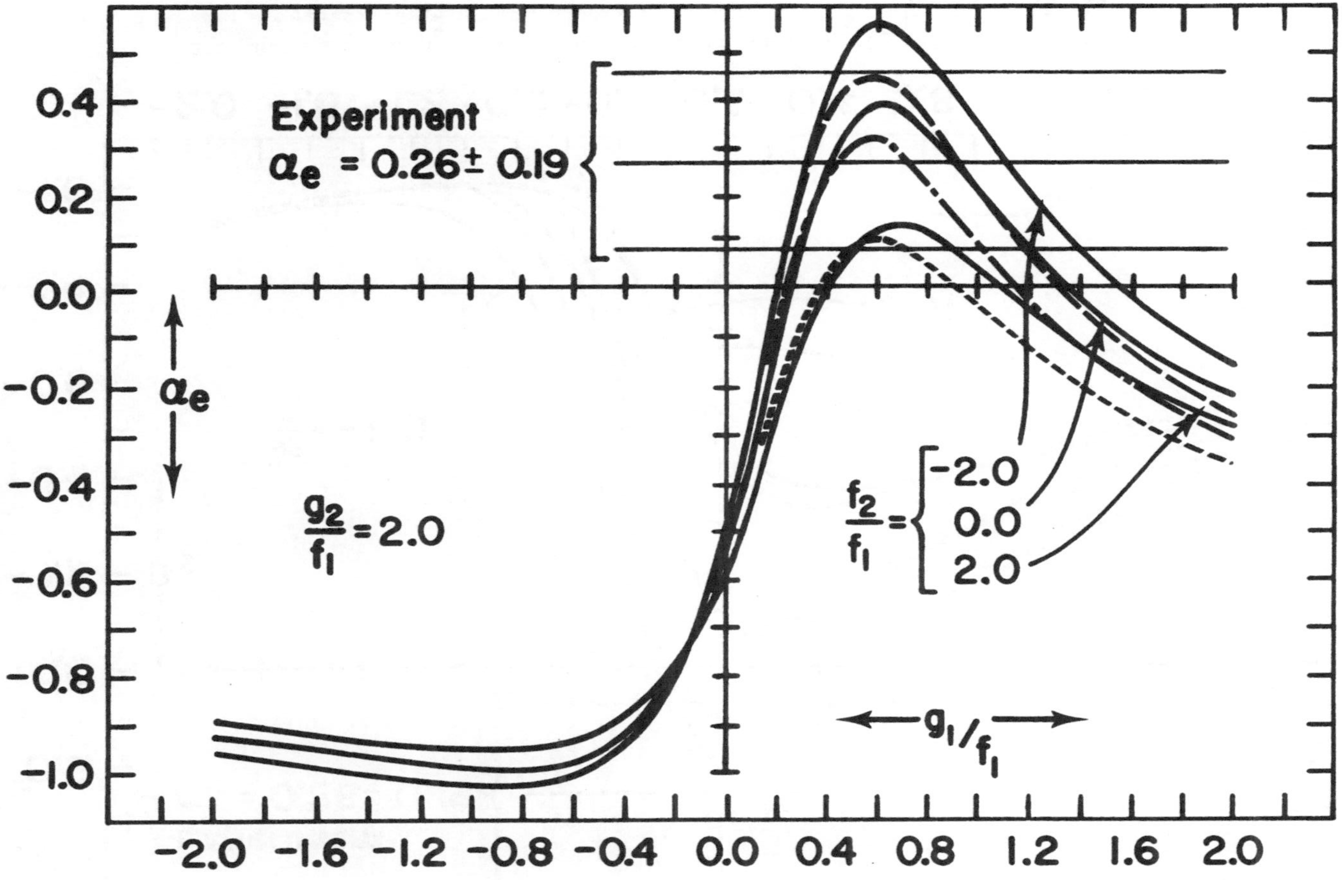

FIG. 5. Plots of α_e in $\Sigma^- \rightarrow ne\nu$ as a function of g_1/f_1 and f_2/f_1. g_2 is kept fixed at 2.0. The dotted lines have the same meaning as in Fig. 1.

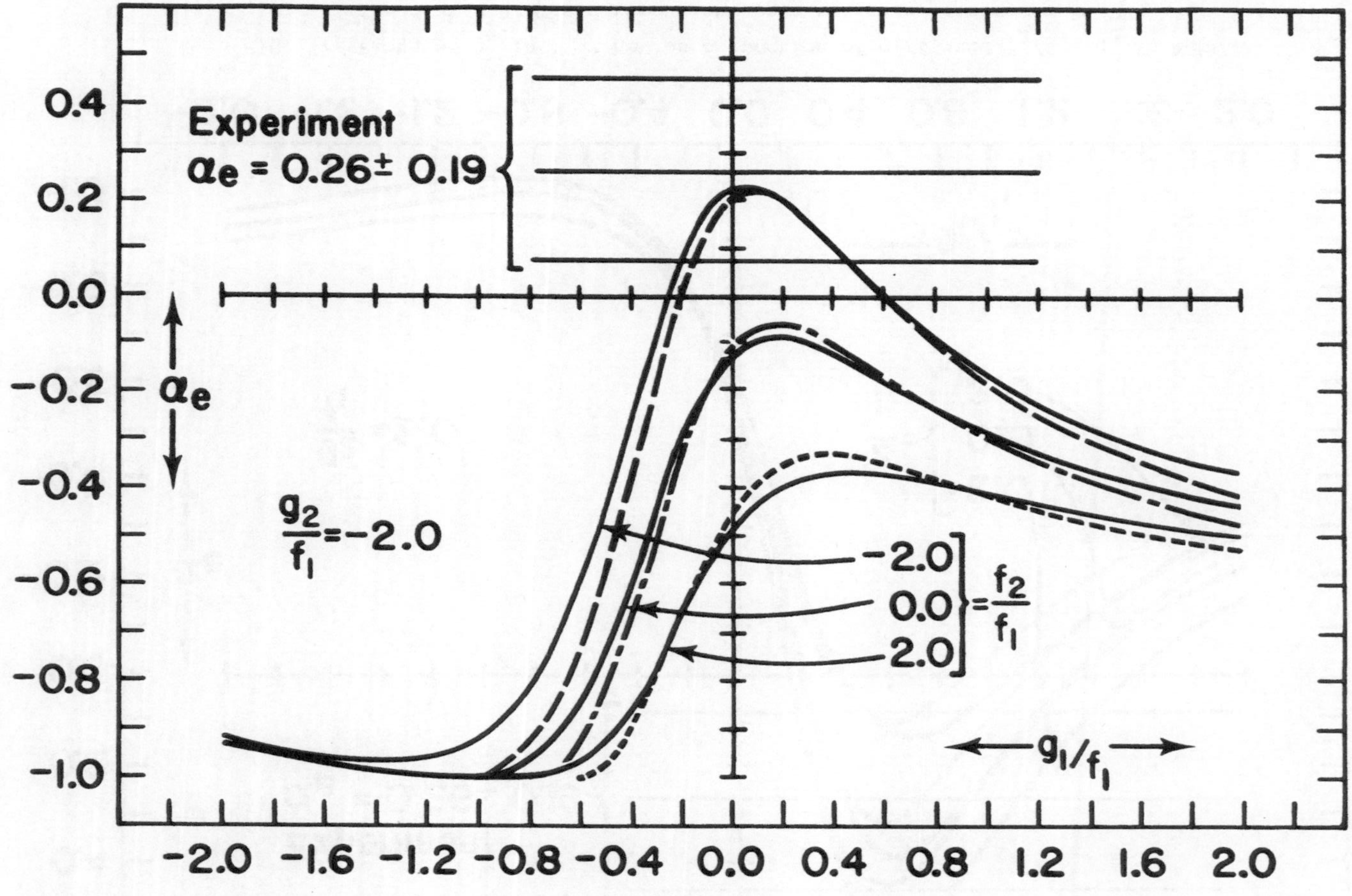

FIG. 6. Plots of α_e in $\Sigma^- \to ne\nu$ as a function of g_1/f_1 and f_2/f_1. g_2 is kept fixed at -2.0. The dotted lines have the same meaning as in Fig. 1.

plots for $\alpha_{e\nu}$, α_e, α_ν, and α_p of $\Lambda \to pe\nu$ can be found in Ref. (2); there is no need to reproduce them here. We have marked on Figs. (1-6) the current experimental values for the corresponding observables, which we borrowed from the next section.

$\alpha_{e\nu}$ in $\Sigma^- \to ne\nu$ is only sensitive to f_2/f_1 near its maximum. Going through Figs. (1-3), one can see that it is quite sensitive to g_2/f_1, in general. It shifts noticeably to the right or to the left, according to this ratio being positive or negative. At any rate, its sensitivity to f_2/f_1 is almost non-existent given its present experimental value even for large values of g_2/f_1. In contrast, $\alpha_{e\nu}$ is a steep function of g_1/f_1. Clearly, the values of g_1/f_1 and g_2/f_1 are strongly correlated to one another in $\alpha_{e\nu}$. The dotted lines are the changes in $\alpha_{e\nu}$ when vector meson dominance estimates for the slope parameters λ_1^f and λ_1^g are included in $\alpha_{e\nu}$. One can see that, kinematically, the q^2-dependence of f_1 and g_1 makes a more important contribution than the f_2/f_1 ratio itself. $\alpha_{e\nu}$ can help to determine accurately g_1/f_1 provided a good estimate or independent experimental information on g_2/f_1 is available. Contrary to what is usually stated, $\alpha_{e\nu}$ provides information on the sign of g_1/f_1 as long as $g_2 \neq 0$. It is only when $g_2 = 0$ that the information on the sign of g_1/f_1 is lost.

Looking at Figs. (4-6), one can see that α_e in $\Sigma^- \to ne\nu$ is again sensitive to f_2/f_1 near its maximum, although it is more sensitive there than $\alpha_{e\nu}$. It looses its sensitivity to f_2/f_1 if its experimental value becomes large and negative, unless g_2/f_1 becomes large and negative too. α_e is quite sensitive to g_2/f_1, and is a very steep function of g_1/f_1 and depends strongly on its sign. The q^2-dependence of f_1 and g_1 is noticeable, but not very important, as is shown by the dotted lines. The behaviour of $\alpha_{e\nu}$ and α_e is similar for other processes with large ΔM. It is clear that in α_e there is a strong correlation between the three form factor ratios, and not only between two of them, as was the case in $\alpha_{e\nu}$.

We shall not pursue further the analysis of this kind of graphs. We refer the reader to Ref. 2. We shall only mention that α_ν of $\Lambda \to pe\nu$, although it is a steep function of g_1/f_1, becomes insensitive to this ratio if the experimental value of α_ν is close to one. In general, α_ν does not show a very important dependence on f_2/f_1; but it does on g_2/f_1, even for experimental values close to one. The role of α_ν in $\Lambda \to pe\nu$ is, in a loose manner of speaking, a rather negative one. If its experimental value were not to be found close to one but somewhat lower, say around 0.75 or 0.80, then a substantial change in g_1/f_1 from large positive values is required to occur.

From the above discussion, it is important then to keep in mind that measuring form factors through integrated observables is a subtle task. One quantity may not show an apparently important deviation from its theoretical prediction and it, nevertheless, may imply a very strong change in some of the form factor ratios. A 10 or 20% deviation in one integrated observable may mean a 60 or 80% change in g_1/f_1 or even more in f_2/f_1 or g_2/f_1. This is the case of α_ν in $\Lambda \rightarrow pe\nu$. While, for other integrated observables it may be required a substantial change from the theoretical prediction to observe a high change in g_1/f_1. This is the case of α_e in $\Sigma^- \rightarrow ne\nu$. One can see also that it is important to watch for the correlation between one form factor ratio and another in the integrated observables. Finally, let us remark that if more than a qualitative graphical analysis is desired then the formulas of Appendix 1 should be used.

3.3. Experimental Data.

The available[7-9] experimental evidence on HSD consists of decay rates, e-ν angular distributions (or equivalently, energy spectra of the emitted hyperon), and, in some cases, angular distribution of an emitted i-th particle with respect to the polarization direction of the decaying or emitted hyperon. The information on angular distributions can be best put into the form of $\alpha_{e\nu}$, α_i, A and B coefficients. In addition, it is customary to quote experimental values for the g_1/f_1 ratios. Some upper bounds on certain transition rates are also available. We have collected all this information in Tables 1, 2, 3, and 4. The first three tables contain the world averages last reported[7], and Table 4 includes the results of two very recent experiments[8,9].

It is somewhat disappointing to realize that of the 16 e-mode decay and 10 μ-mode decays only 7 e-mode decays and 3 μ-modes decays have been observed. The last four decays of Table 2.1 —or last 8 including the μ-modes— are not expected to exist (see next chapter) and only not very meaningful upper bounds have been set on one of them —or two including the corresponding μ-mode. The angular information is even scarcer.

It is complete only for $n \to pe\nu$, appreciable for $\Lambda \to pe\nu$ and $\Sigma^- \to ne\nu$, but insufficient or non-existent for the other observed decays. The accumulation of information in Tables 1-4 covers more than two decades, involving the effort of many people. There is a clear mismatch between the progress of experiment and the progress of theory in HSD. It is only until very recently that the bulk of Tables 1-4 has been collected. To give oneself an idea it should be sufficient to mention that until 1975 the V-A theory was not even required to describe $\Sigma^- \to ne\nu$, only a V or an A interaction would have sufficed for it. Nevertheless, as we shall see in the next chapters this experimental evidence will lead to meaningful comparisons with our current theoretical beliefs.

The comparison of theory with the g_1/f_1 ratios can lead to inconsistencies when the q^2-dependence of these form factors is considered. The problem is quite subtle. The world averages for $(g_1/f_1)_{\Lambda p}$ and $(g_1/f_1)_{\Sigma^- n}$ quoted in Table 2 were obtained averaging experimental values obtained under different assumptions; in some cases no q^2-dependence for $g_1(q^2)$ and $f_1(q^2)$ was taken into account, and in other cases a particular choice for their q^2-dependence was made. This should not present any problem as long as the experimental error bars on these ratios are substantially larger than the uncertainty introduced by the q^2-dependence contributions, as is the case in neutron decay. But in $\Lambda \to pe\nu$ and $\Sigma^- \to ne\nu$ the error bars are comparable and even smaller than such an uncertainty. This consistency problem can be avoided by using correlation and asymmetry coefficients, since their values do not involve any assumption about the q^2-dependence. A similar discussion applies when the contributions of the induced form factors are considered.

There are nevertheless several points about the angular coefficients that must be mentioned. In some experiments only values for the absolute value of the g_1/f_1 ratios are given and the values of the $\alpha_{e\nu}$ coefficients are not quoted. The corresponding $\alpha_{e\nu}$ values can be obtained by substituting the $|g_1/f_1|$ into the general expressions for the $\alpha_{e\nu}$ under exactly the same assumptions that were used in obtaining the $|g_1/f_1|$. This is the case of $\alpha_{e\nu}^{\Sigma^-\Lambda}$ and $\alpha_{e\nu}^{\Sigma^- n}$. In addition, for $\alpha_{e\nu}^{\Sigma^- n}$ there are two experimental results available,[7] but they differ too much and we have chosen the result of Ref. (10) only, for the compilations of Tables 1 and 2. Our choice is based on a very recent result[9] that confirms the measurement of Ref. (10). These new results are incorporated in Table 4. (There the new $\Xi^- \to \Sigma^0 e\nu$ rate replaces the $\Xi^- \to \Lambda,\Sigma^0 e\nu$ rate).

An objection may be raised to using $\alpha_e^{\Lambda p}$, $\alpha_\nu^{\Lambda p}$, and $\alpha_p^{\Lambda p}$, since their experimental values are not statistically independent and there is a risk of introducing some bias. We can appreciate the importance of this bias by fitting $(g_1/f_1)_{\Lambda p}$ to these asymmetries alone and comparing the result to the value of $(g_1/f_1)_{\Lambda p}$ obtained from the statistically independent combinations $\alpha_e^{\Lambda p} + \alpha_\nu^{\Lambda p}$, $\alpha_e^{\Lambda p} + \alpha_p^{\Lambda p}$, and $\alpha_\nu^{\Lambda p} + \alpha_p^{\Lambda p}$, using in both cases the same assumptions. The value from our fit is 0.32 ± 0.10 and from Ref. (11) it is $0.33 {+0.14 \atop -0.09}$. There is, indeed, some bias, but it is much smaller than the corresponding error bars. So, for the meantime we can safely use each asymmetry coefficient in $\Lambda \to pe\nu$. Nevertheless, this point should be kept in mind and it is recommended that in future experiments values for the statistically independent combinations should also be given.

We shall use the available evidence on free neutron decay. Often the vector coupling constant obtained in superallowed nuclear decays[12] is used. This is of course quite all right. But given the recent improvements in the free neutron decay measurements, we find it more attractive to limit ourselves to using data on elementary-particle decays. As a matter of fact, it is most desirable that no information from other fields be used. This way the value of the vector coupling constant coming from particle physics can be compared to the one coming from nuclear physics, thus providing some independent check on the accuracy of the nuclear-structure and isospin-breaking calculations used in studying superallowed decays.

Safe as it seems, converting a g_1/f_1 ratio to an $\alpha_{e\nu}$, when this last was not quoted by the experimental groups themselves, may carry another subtle problem. Ideally, the procedure we have described above is expected to work well. In practice, apparata have acceptance limitations. How to take these limitations into account is an open question, and so it is not clear how important a bias may be introduced by neglecting them, as we have done. However, we are looking for big effects, that amount to several standard deviations, and in as much as experiments are low in statistics the error we may be introducing is expected to be less appreciable than the effects we shall study in the next chapters. We can only apologize for this and encourage experimentalists to quote acceptance independent numbers, in addition to the model dependent g_1/f_1 ratios that are so often quoted. In consequence, we have put in Table 4 statistical world averages only.

Finally, we want to call attention to the fact that we shall always work within the V-A theory, as we implied in Chapter 2. There[13] remains the question as to whether this theory is exactly true or if other forms of the weak interaction — S, T, or P or simple V+A — coexist to greater or lesser extent with the V-A theory. The best measured decay, $n \to pe\nu$, does not yet exclude other forms down to safe small contributions, if at all. In Chapter 10 we shall return to this question.

TABLE 1. Experimental data on HSD corresponding to the world averages last reported[7]. $\alpha_{e\nu}$ in $\Sigma^{\pm} \to \Lambda e\nu$ and $\Sigma^{-} \to ne\nu$ are computed by us, as explained in the text, and their error bars are only statistical. $\alpha_{e\nu}$ in $\Lambda \to pe\nu$ of Wise et al.[7] is also computed by us, and afterwards it is averaged into the world average quoted by Lindquist et al.[11] The $n \to pe\nu$ rate is in 10^{-3} sec^{-1}, all other rates are in 10^{6} sec^{-1}.

Process	Experimental value
$n \to pe\nu$ (rate)	1.081 ± 0.013
$\Sigma^{+} \to \Lambda e\nu$ (rate)	0.250 ± 0.063
$\Sigma^{-} \to \Lambda e\nu$ (rate)	0.412 ± 0.034
$\Lambda \to pe\nu$ (rate)	3.172 ± 0.062
$\Sigma^{-} \to ne\nu$ (rate)	7.29 ± 0.28
$\Xi^{-} \to \Lambda e\nu$ (rate)	1.77 ± 0.67
$\Xi^{-} \to \Lambda,\Sigma_{o} e\nu$ (rate)	4.14 ± 0.13
$\Lambda \to p\mu\nu$ (rate)	0.60 ± 0.13
$\Sigma^{-} \to n\mu\nu$ (rate)	3.04 ± 0.27
$\Xi^{-} \to \Lambda\mu\nu$ (rate)	2.13 ± 2.13
$n \to pe\nu$ ($\alpha_{e\nu}$)	-0.074 ± 0.004
$n \to pe\nu$ (α_{e})	-0.083 ± 0.002
$n \to pe\nu$ (α_{ν})	0.998 ± 0.025
$\Sigma^{\pm} \to \Lambda e\nu$ ($\alpha_{e\nu}$)	-0.35 ± 0.15
$\Sigma^{-} \to \Lambda e\nu$ ($\alpha_{e\nu}$)	-0.36 ± 0.11
$\Sigma^{-} \to ne\nu$ ($\alpha_{e\nu}$)	0.28 ± 0.05
$\Sigma^{-} \to ne\nu$ (α_{e})	0.04 ± 0.27
$\Lambda \to pe\nu$ ($\alpha_{e\nu}$)	-0.018 ± 0.018
$\Lambda \to pe\nu$ (α_{e})	0.125 ± 0.066
$\Lambda \to pe\nu$ (α_{ν})	0.821 ± 0.060
$\Lambda \to pe\nu$ (α_{p})	-0.508 ± 0.065

TABLE 2. Angular experimental data of HSD in the form of g_1/f_1 ratios. The sources are in Ref. (7).

$(g_1/f_1)_{np}$	1.255 ± 0.006
$(g_1/f_1)_{\Lambda p}$	0.690 ± 0.034
$\lvert g_1/f_1 \rvert_{\Sigma^{-}n}$	0.435 ± 0.035
$(f_1/g_1)_{\Sigma^{-}\Lambda}$	-0.14 ± 0.24

TABLE 3. Upper-bounds for the branching ratios of forbidden or allowed but as yet unobserved decays. The sources are in Ref. (7).

$\Sigma^+ \to ne\nu$	$< 5 \times 10^{-6}$
$\Sigma^+ \to n\mu\nu$	$< 3 \times 10^{-5}$
$\Xi^0 \to pe\nu$	$< 1.3 \times 10^{-3}$
$\Xi^0 \to p\mu\nu$	$< 1.3 \times 10^{-3}$
$\Xi^- \to ne\nu$	$< 3.2 \times 10^{-3}$
$\Xi^- \to n\mu\nu$	$< 1.5\%$
$\Xi^0 \to \Sigma^- e\nu$	$< 0.9 \times 10^{-3}$
$\Xi^0 \to \Sigma^- \mu\nu$	$< 0.9 \times 10^{-3}$
$\Xi^0 \to \Sigma^+ e\nu$	$< 1.1 \times 10^{-3}$
$\Xi^0 \to \Sigma^+ \mu\nu$	$< 1.1 \times 10^{-3}$
$\Xi^- \to \Sigma^0 \mu\nu$	$< 8 \times 10^{-4}$
$\Xi^- \to \Xi^0 e\nu$	$< 2.3 \times 10^{-3}$

TABLE 4. Experimental data on HSD incorporating into Table 1 the results of several recent experiments. Units are as in Table 1. The upper index a means that the sources are References (7) and (9), and the upper index b means that the source is Reference (8). The angular coefficients of the last four lines are computed by us.

Process	Experimental value
$n \to pe\nu$ (rate)	1.081 ± 0.013
$\Sigma^+ \to \Lambda e\nu$ (rate)	0.250 ± 0.063
$\Sigma^- \to \Lambda e\nu$ (rate)[a]	0.388 ± 0.018
$\Lambda \to pe\nu$ (rate)[a]	3.186 ± 0.057
$\Sigma^- \to ne\nu$ (rate)[a]	6.96 ± 0.22
$\Xi^- \to \Lambda e\nu$ (rate)[a]	3.31 ± 0.18
$\Xi^- \to \Sigma^o e\nu$ (rate)[a]	0.53 ± 0.10
$\Lambda \to p\mu\nu$ (rate)	0.6 ± 0.13
$\Sigma^- \to n\mu\nu$ (rate)	3.04 ± 0.27
$\Xi^- \to \Lambda\mu\nu$ (rate)	2.13 ± 2.13
$n \to pe\nu$ ($\alpha_{e\nu}$)	-0.074 ± 0.004
$n \to pe\nu$ (α_e)	-0.083 ± 0.002
$n \to pe\nu$ (α_ν)	0.998 ± 0.025
$\Sigma^\pm \to \Lambda e\nu$ ($\alpha_{e\nu}$)	-0.35 ± 0.15
$\Sigma^- \to \Lambda e\nu$ ($\alpha_{e\nu}$)[a]	-0.404 ± 0.044
$\Sigma^- \to ne\nu$ ($\alpha_{e\nu}$)[a]	0.279 ± 0.026
$\Sigma^- \to ne\nu$ (α_e)[b]	0.26 ± 0.19
$\Lambda \to pe\nu$ ($\alpha_{e\nu}$)[a]	-0.013 ± 0.014
$\Lambda \to pe\nu$ (α_e)	0.125 ± 0.066
$\Lambda \to pe\nu$ (α_ν)	0.821 ± 0.060
$\Lambda \to pe\nu$ (α_p)	-0.508 ± 0.065
$\Xi^- \to \Lambda e\nu$ ($\alpha_{e\nu}$)[a]	0.53 ± 0.10
$\Xi^- \to \Lambda e\nu$ (A)[a]	0.62 ± 0.10
$\Sigma^- \to \Lambda e\nu$ (A)[a]	0.07 ± 0.07
$\Sigma^- \to \Lambda e\nu$ (B)[a]	0.85 ± 0.07

Chapter 4. The Cabibbo Theory

4.1. Selection Rules.

Even if weak interactions violate many conservation laws of strong interactions —like strangeness, isospin or parity— one may still expect that they reflect somehow the existence of the corresponding conserved quantum numbers. To say it in a few words, even though weak interactions may not be invariant under a certain symmetry transofrmation, they may still be covariant under it. A classical example of this situation is the role parity plays in the leptonic decays of spin zero pseudoscalar mesons. In $K \to \mu\nu$ or in $\pi \to \mu\nu$ there is parity violation, but only the axial-vector current A_μ can participate in the hadronic part of the transition amplitude, because K, π, and the hadronic vacuum have well defined parities and that excludes one of the currents $-V_\mu$ in this case. Clearly covariance properties can still play a role, despite there is no longer invariance.

With respect to internal symmetries of strong and electromagnetic interactions, one may expect that weak interactions also exhibit well defined covariance properties[1.2]. The $\Delta S = \Delta Q$ rule is a very encouraging example. If the hadronic weak current that mediates a decay where $\Delta S \neq 0$ is denoted by $j^{\Delta S}$, then,we know that its commutator with the charge operator Q is

$$[j_\mu^{\Delta S}, Q] = j_\mu^{\Delta S}, \tag{1}$$

when $\Delta Q = 1$. One may then propose that the commutator with the strangeness operator S be

$$[j_\mu^{\Delta S}, S] = j_\mu^{\Delta S}, \tag{2}$$

in analogy with Eq. (1). It then follows that the two commutators must be equal to one another, and thus $\Delta S = \Delta Q$. If the opposite sign had been chosen in Eq. (2), then $\Delta S = -\Delta Q$ would have followed. Also, Eqs. (1) and (2) require that there exist no HSD with $|\Delta S| > 2$.

The evidence for the choice in Eq. (2) comes from Table 3.3. The decay $\Sigma^+ \to ne\nu$

has been looked for and has not yet been found. The upper bound on its branching ratio is not yet quite stringent but it is nevertheless very encouraging. Decays with $|\Delta S| > 2$ have been looked for too. Their upper bounds are not too stringer either yet, but one can at least admit that such decays are certainly not abundant. It is therefore reasonable to accept the existence of covariance properties like Eq. (2) in HSD. The $\Delta S = \Delta Q$ rule applies, of course, only when $\Delta S \neq 0$.

This rule can be given another form using the Gell-Mann-Nishijima rule[1.1],

$$Q = I_3 + \frac{1}{2}\,(S+B), \tag{3}$$

where I_3 is the third component of isospin and B is the baryon number. Since B is always conserved, it follows from Eq. (3) and $\Delta S = \Delta Q$, that $\Delta I_3 = \pm 1/2$ according to the sign of S. From this rule one can infer that $j_\mu^{\Delta S}$ should transform like an insodoublet. If $\Delta S = -\Delta Q$ were to occur then $\Delta I = 3/2$ would be permitted, and $j_\mu^{\Delta S}$ would transform as a higher rank isotensor.

Another selection rule in weak decays is the G-parity rule we mentioned in Section 2.2. The evidence for its validity comes mainly from nuclear β decay. Currently one can assess that if second-class A_μ currents exist, then their contribution to nuclear β decay is much suppressed[3.12]. For V_μ currents no second-class currents must exist if they are to share properties with the electromagnetic current, which is known to be first-class only. The conserved vector current hypothesis (CVC) states what is currently accepted as the proper connection between V_μ and the electromagnetic current.

The full realization of the relevance of covariance properties of weak currents was materialized in the Cabibbo theory[1], to which we now turn.

4.2. Postulates of the Cabibbo theory.

The Cabibbo theory[1] (CT) contains several postulates that emerged gradually over a period of six to eight years, until they were put together in the early 60's by Cabibbo in an elegant formulation —which illustrates in a splendid fashion how generalizations can be made in weak interactions.

The postulates are

1) Only first-class hadronic operators participate in weak interactions.

2) The V-A theory is valid.

3) The V_μ and A_μ currents are octet SU(3) operators.

4) V_μ is related to the electromagnetic current through the CVC hypothesis.

5) The universality of weak interactions is determined by putting

$$G_v = G_\mu \cos\theta$$

for $\Delta S = 0$ decays, and

$$G_v = G_\mu \sin\theta$$

for $|\Delta S| = 1$ decays.

6) The SU(3) symmetry limit is valid, except that the hadron mass differences should be retained.

The order in which we have listed the postulates of the CT is intended to reflect to a certain extent their importance. With the advent of renormalizable gauge theories it has become clear that postulate 1 implies great simplicity. Nobody has really proved that the existence of second-class currents would invalidate the renormalizability of gauge theories in general, but it is quite certain that their very attractive simplicity would be lost. Postulates 2, 3, and 4 should follow this order. V_μ may be an SU(3) octet and may not be necessarily related to the electromagnetic current. The universality formulation was assumed to be very general at the time of its introduction, but as we shall see in Sec. 8.4 it may require changes when new flavors are considered. The weakest of all the postulates is number 6. In this respect it should be well kept in mind that the CT has never been intended to be exact. As a matter of fact, since we know that SU(3) is a broken symmetry in the real world, one must eventually find discrepancies between CT and experiment.

There are other assumptions in the CT which are not expected to have any fundamental character. These are simply working assumptions. Nothing is said about the problem of radiative corrections, nor about the q^2-dependence of the form factors.

or if g_3 is related to g_1 in any way. One must certainly keep track of these working assumptions before one jumps to conclude whether the CT agrees well or badly with experiment. In contrast, the assumption about the smallness of SU(3) breaking is more than just a working hypothesis, because the usefulness of an internal symmetry is intimately related to the extent to which it is approximately exact.

We shall not pursue further the critique of CT for a while. It is better to appreciate first its predictive power and to compare it with experiment. The critique should then be tuned to the results we find.

4.3. Predictions for Form Factors and Undetermined Parameters.

Within CT all of the decays in Table 2.1 become related to one another. In the 16 e-mode processes we have 64 unknown form factors. The CT determines all of them in terms of a few quantities —7 quantities in all, as we shall see rightaway. Thus, CT has quite some predictive power in HSD.

Because of postulates 1 and 6, the pseudotensor form factors g_2 are all predicted to be zero. Postulate 3 implies immediately that the $\Delta S = \Delta Q$ rule is valid, predicting then that the last four decays of Table 2.1 do not exist. It also implies along with postulate 6 that the non-vanishing vector and axial-vector form factors are of the form

$$f_i^{AB}(0) = C_F^{AB} F_i + C_D^{AB} D_i, \quad i = 1,2 \tag{4}$$

and

$$g_1^{AB}(0) = C_F^{AB} F + C_D^{AB} D. \tag{5}$$

Here C_F^{AB} and C_D^{AB} are SU(3) Clebsch-Gordan coefficients that appear when an octet operator is "sandwiched" between octet states; the upper indeces refer to the hadrons involved and the lower indeces F and D correspond to the antisymmetric and symmetric octets that appear in the direct product of two octets. These coeffients are given in Table 1 for the 7 observed decays of Table 2.1. Our conventions are those of Ref. (2).

TABLE 1. Clebsch-Gordan coefficients relevant to observed HSD

	C_F^{AB}	C_D^{AB}
$n \to p$	$\frac{1}{\sqrt{6}}$	$-\left(\frac{3}{10}\right)^{1/2}$
$\Sigma^{\pm} \to \Lambda$	0	$-\frac{1}{\sqrt{5}}$
$\Lambda \to p$	$-\frac{1}{2}$	$\frac{1}{2\sqrt{5}}$
$\Sigma^- \to n$	$-\frac{1}{\sqrt{6}}$	$-\left(\frac{3}{10}\right)^{1/2}$
$\Xi^- \to \Lambda$	$\frac{1}{2}$	$\frac{1}{2\sqrt{5}}$
$\Xi^- \to \Sigma^0$	$\frac{1}{2\sqrt{3}}$	$-\frac{1}{2}\left(\frac{3}{5}\right)^{1/2}$

The constants F_i, D_i, F, and D, i = 1,2, are reduced form factors. These 6 quantities together with the Cabibbo angle θ, introduced in postulate 5, make the 7 quantities we referred to above. The first four can be determined using the CVC hypothesis, postulate 4, together with the symmetry limit, postulate 6. f_1 and f_2 are thus related to the electromagnetic form factors of the nucleons that have been pretty well measured. In this way, f_1 is related to the charges of the neutron and the proton; one gets[2]

$$F_1 = \sqrt{6} , \tag{6}$$

and

$$D_1 = 0 . \tag{7}$$

The connection of f_2 with the magnetic moments of the nucleons has a subtlety. When we defined the form factors of HSD in Eq. (2.12) we divided f_2 by M_1, in order that it have the same units as f_1. The corresponding f_2 in the nucleon electromagnetic matrix-element is divided by the proton mass[(3)]. In as much as the symmetry limit is assumed to be valid, we must retain this convention and replace in Eq. (2.12)

$$f_2 \rightarrow \frac{M_1}{M_p} f_2 ,$$

Then it is this latter f_2 that has the form given in Eq. (4). Thus, we get

$$F_2 = \left(\frac{\mu_p}{2} + \frac{\mu_n}{4} \right) \sqrt{6} , \qquad (8)$$

and

$$D_2 = \frac{\mu_n}{4} \sqrt{30} . \qquad (9)$$

where μ_p and μ_n are the anomalous magnetic moments of the proton and the neutron, respectively. Their current values[3.7] are

$$\mu_p = 1.79285,$$

$$\mu_n = -1.91304.$$

There remain three undetermined parameters F, D and θ. They cannot be brought into HSD from experimental evidence in other fields of Particle Physics, although if θ is common to mesons and hyperons one could borrow the value it has in meson leptonic decays. This is, however, not very reliable and accordingly we shall keep it as a free parameter. The ratio of F and D could be fixed by making appeal to a higher symmetry like SU(6), but this would not be reliable either. We shall then keep F, D, and θ as three undetermined parameters to be fixed within HSD. The values obtained for them would serve to check on their values obtained in other schemes or decays.

The overall coupling constant G_μ comes from μ-decay. After applying radiative corrections to the μ lifetime, its current value is given in Eq. (2.6).

4.4. Comparison with Experiment (Rough).

As a starting point, we have performed a rough comparison of the CT with the data of Table 3.1 ignoring the working assumptions we mentioned in Sec. 2. That is, we shall put the q^2-dependence of f_1 and g_1 equal to zero, ignore radiative corrections, and, for the μ-mode decays, ignore the presence of g_3. Our aim is not quite to test the CT, but rather to develop a feeling for the quality of the data of Tables 3.1 and 3.2. The upper bounds of Table 3.3 are automatically satisfied in the CT.

The results are displayed in Table 2. The second column gives the prediction for each observable of Table 3.1. The last column gives the contribution to χ^2 of each prediction. The quantities marked with an asterisk were not used in the fit. The values of the three free parameters are in the last lines of this table, as well as the total value of χ^2. The number of degrees of freedom n_D is 18.

From the χ^2 point of view the fit is poor, with $\chi^2 = 44.3$. The large value of χ^2 is built up essentially by six quantities: $n \to pe\nu$ (rate), $\Sigma^- \to \Lambda e\nu$ (rate), $\Xi^- \to \Lambda e\nu$ (rate), $\Lambda \to pe\nu$ (α_e and α_ν), and $\Sigma^- \to ne\nu$ (α_e). The contributions of the first, fifth and last quantities are big, each amounts to more than 2.5 standard deviations.

Although, we have not fitted the g_1/f_1 ratios, it is interesting to remark that the total χ^2 corresponding to them is $\chi^2 = 6.9$, made up essentially by g_1/f_1 of $n \to pe\nu$. This shows to what extent their experimental values are insensitive. If these were the only pieces of data avalilable to us, instead of the angular coefficients, we would be led to conclude that the CT is in very good agreement with experiment. It is clear that using the angular coefficients, instead of the g_1/f_1 ratios, not only avoids inconsistencies but provides a more sensitive test. From this point of view we can conclude that, although HSD decay data are still rather meager, they are already restrictive enough to force us to look into the CT assumption with more care.

It will be our task in the next four chapters to revise the CT critically. We shall begin by questioning the working assumptions we pointed out at the end of Sec. 2.

TABLE 2. Rough comparison of the CT with the HSD data of Table 3.1. No radiative corrections, nor q^2-dependence of the leading form factors are accounted for. The asterisk on some quantities means that they were not fitted; their quoted $\Delta\chi^2$ indicates the contribution to χ^2 the predictions for them would make — the data of Table 3.2 was used.

	Prediction	$\Delta\chi^2$
$n \to pe\nu$ (rate)	1.045	7.88
$\Sigma^+ \to \Lambda e\nu$ (rate)	0.288	0.35
$\Sigma^- \to \Lambda e\nu$ (rate)	0.476	3.56
$\Lambda \to pe\nu$ (rate)	3.178	0.01
$\Sigma^- \to ne\nu$ (rate)	7.043	0.78
$\Xi^- \to \Lambda e\nu$ (rate)	2.822	2.46
$\Xi^- \to \Lambda,\Sigma^0 e\nu$ (rate)	3.363	0.36
$\Lambda \to p\mu\nu$ (rate)	0.510	0.48
$\Sigma^- \to n\mu\nu$ (rate)	3.124	0.10
$\Xi^- \to \Lambda\mu\nu$ (rate)	0.764	0.41
$n \to pe\nu$ ($\alpha_{e\nu}$)	-0.078	1.17
$n \to pe\nu$ (α_e)	-0.087	3.16
$n \to pe\nu$ (α_ν)	0.988	0.17
$\Sigma^\pm \to \Lambda e\nu$ ($\alpha_{e\nu}$)	-0.400	0.11
$\Sigma^- \to \Lambda e\nu$ ($\alpha_{e\nu}$)	-0.403	0.15
$\Sigma^- \to ne\nu$ ($\alpha_{e\nu}$)	0.347	1.77
$\Sigma^- \to ne\nu$ (α_e)	-0.681	7.13
$\Lambda \to pe\nu$ ($\alpha_{e\nu}$)	0.016	3.62
$\Lambda \to pe\nu$ (α_e)	0.017	2.69
$\Lambda \to pe\nu$ (α_ν)	0.977	6.78
$\Lambda \to pe\nu$ (α_e)	-0.579	1.19
$(g_1/f_1)_{np}$	1.268*	4.47
$(g_1/f_1)_{\Lambda p}$	0.715*	0.54
$\lvert g_1/f_1\rvert_{\Sigma^- n}$	$\lvert$-0.390*$\rvert$	1.65
$(f_1/g_1)_{\Sigma^-\Lambda}$	0.000*	0.34

$F = 1.075 \pm 0.019$ $\qquad$ $D = -1.513 \pm 0.016$

$\sin\theta = 0.233 \pm 0.002$ $\qquad$ $\chi^2 = 44.33$

Chapter 5. Radiative Corrections

5.1. Problems.

The calculation of radiative corrections to processes where hadrons are involved has been a long standing problem. In attempting to compute them one runs into many difficulties. They diverge in the ultraviolet region, they are affected by strong interactions and by the details of the ultimate theory of weak interactions. In addition, in HSD the momentum transfer to the leptons is not negligibly small, and this brings in more complications.

As we have stressed several times before, one of the attractive features of β decays is that they serve as a microscope to study hadrons. This is valid as long as the factorization property of the weak transition amplitude M_o, Eq. (2.11), can be retained. Radiative corrections spoil the resolution of this microscope, so to speak. That is, radiative corrections spoil this factorization, and, therefore, one must compute them and add them to the differential decay rate expressions of Sec. 2.3, if one wishes to restore the ability of the lepton pair to explore clearly the hadrons involved in β decays.

We are thus led into a circular argument. We want β decays to tell us something about strong interactions whose theory we still do not know. But, their ability to do so is spoiled by radiative corrections. We cannot compute them because they are affected by strong interactions. This circle repeats itself for the details of weak interactions. If we want to go beyond an effective V-A formulation of β decay we find that radiative corrections again depend on those details, and, thus, they can only be computed in a model dependent fashion by choosing a certain model of weak interactions. Nevertheless, the situation is not as bad as it may seem. There is a way to open these two circles as we shall see in Sec. 3, but we shall have to pay a price for it.

The question of the ultraviolet divergence may no longer be a problem if we believe that gauge theories provide a framework that is general enough and that will contain the ultimate theory of weak interactions. It became customary to deal with the ultraviolet divergence by introducing a cut-off. Gauge theories fix this cut-off by relating it to the mass of the intermediate boson. For our purposes it will only be necessary that the cut-off procedure makes sense, i.e., that the ultraviolet divergence is not present at all and that the terms that contain it can be manipulated as ordinary algebraic quantities.

In addition to the virtual radiative corrections, we must also account for the emission of real photons, i.e., we must add the inner-bremsstrahlung contributions. Real photons accompanying HSD are not discriminated, so we must include their contribution into the differential decay rate. At least soft photons are welcome because with them the infra-red divergence is eliminated.

Although one must be very ambitious in trying to solve a problem, such as the one of radiative corrections, one must also be practical. As a matter of fact, the extent to which one wishes to be rigorous in computing radiative corrections is governed by experimental precision. For the past two decades form factors in HSD have been measured within a precision of 10 to 15%. Currently, experimental precision allows them to be determined to 4 or 5%. We may expect that in this and the next decade they will be measured to 1%. It turns out that terms proportional to q in the radiative corrections to HSD do not amount to more than a fraction of a percent, around 0.3%. It is therefore a good approximation to neglect q in computing the virtual and bremsstrahlung corrections. This approximation will make our life much easier, especially in the bremsstrahlung case. In what follows we shall always use it.

5.2. Model Independent Radiative Corrections.

To first order in α, the fine structure constant, we must consider the Feynman graphs displayed in Fig. 1 for the virtual radiative corrections, and the graphs of Fig. 2 for the bremsstrahlung ones. The blobs in these graphs represent the effects of strong interactions and, at the weak vertex, the effects of details of weak interactions, too.

Some time ago, Sirlin[1] gave a general discussion of the radiative corrections to the energy spectrum of the electron in $n \to pe\nu$. He showed that the virtual radiative corrections can be separated into two parts. One is model independent, finite in the

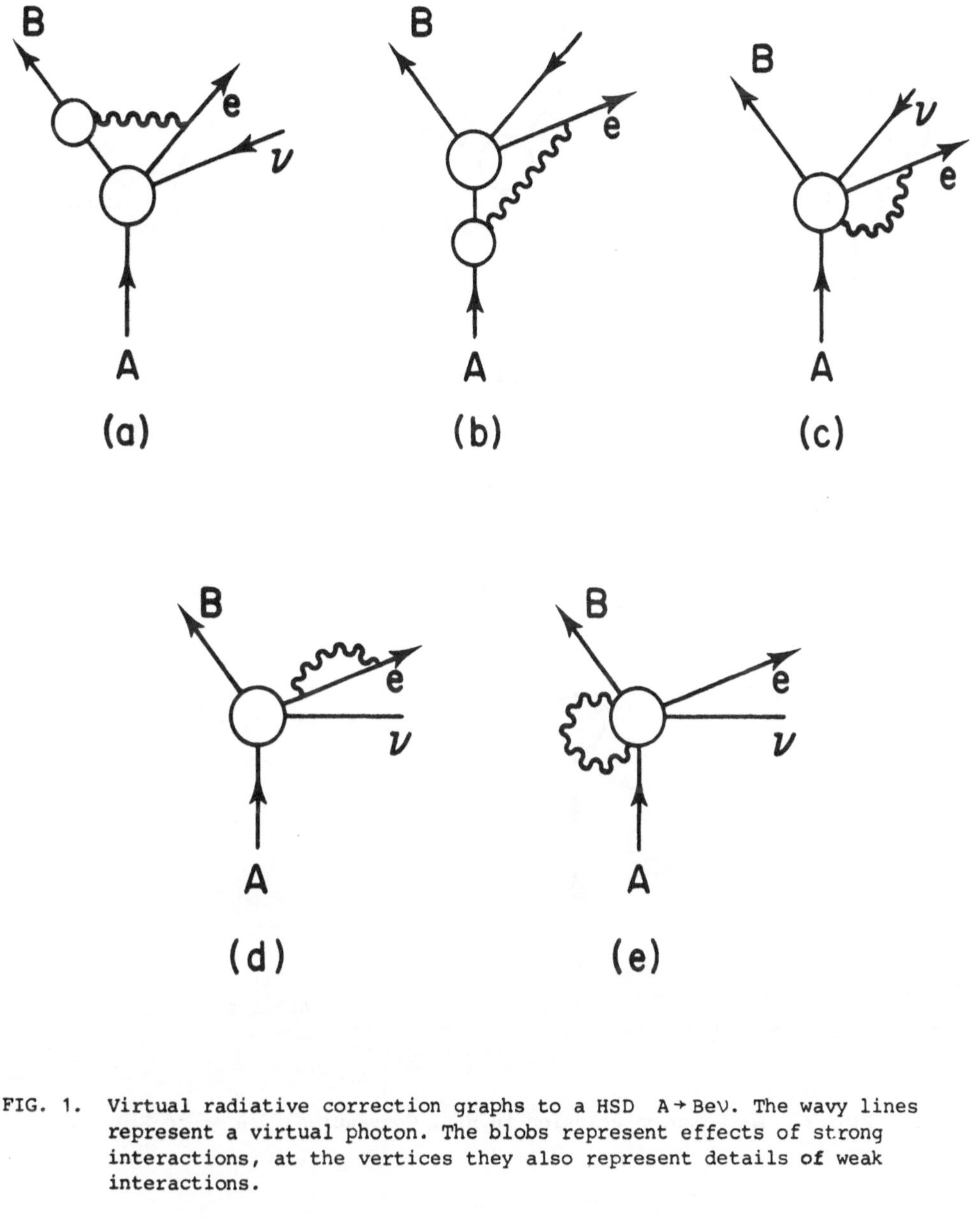

FIG. 1. Virtual radiative correction graphs to a HSD $A \to Be\nu$. The wavy lines represent a virtual photon. The blobs represent effects of strong interactions, at the vertices they also represent details of weak interactions.

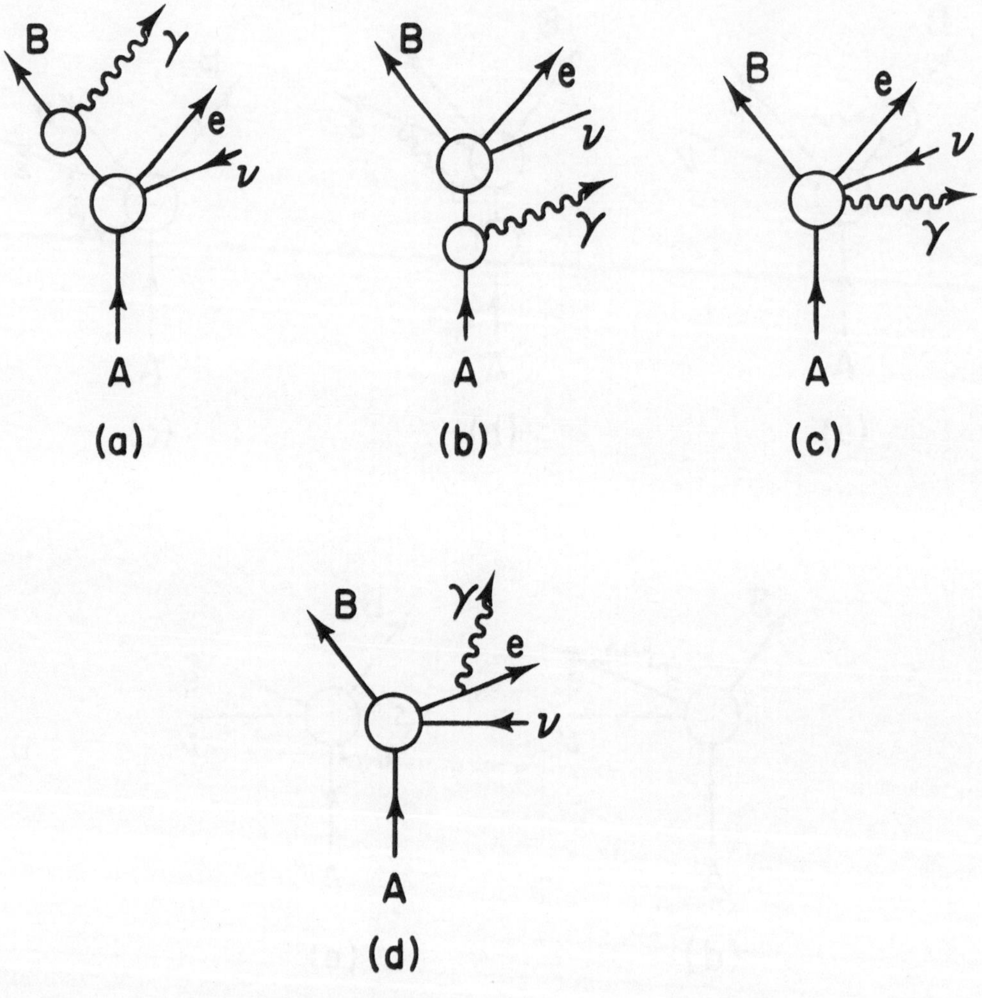

FIG. 2. Real photon emission in HSD. Conventions are as in Fig. 1.

ultraviolat region and contains the infra-red divergence. The other one contains all the complications due to strong interactions and the presence of the intermediate vector boson and the ultraviolet divergence. This separation procedure is gauge invariant. It was later shown[2] that this separation can be generalized to other observables in $n \to pe\nu$ and that it remains valid even when q cannot be neglected. It is therefore applicable to HSD. We shall use this procedure to deal with radiative corrections here[2.7].

The separation relies only on general principles such as Lorentz covariance, analiticity of strong and weak interactions, and, of course, on the validity of quantum electrodynamics. The blobs in Figures 1(a)-1(e) are translated into off-shell electromagnetic form factors of A and B and an electro-weak vertex $W_\lambda(p_1,p_2,k)$; k stands for the four-momentum of the virtual photon. It is not necessary to repeat all the details of the separation steps; it would take several pages to do so, and, hence, it is better to refer the interested reader to the original paper of Sirlin[1] and to Ref. (2). Graphs 1(a)-1(c) contribute to an amplitude M_1,

$$M_1 = -\frac{G_V}{\sqrt{2}}\frac{\alpha}{4\pi^3 i}\int d^4k\, D_{\mu\nu}(k)\,\bar{u}_\ell \frac{2\ell_\nu - \gamma_\nu \not{k}}{k^2 - 2\ell\cdot k + i\varepsilon} O_\lambda v_\nu \times$$

$$\times \left[\bar{u}_B \left\{ \frac{(2p_{1\mu}-k_\mu)W_\lambda(p_1,p_2)}{k^2-2p_1\cdot k + i\varepsilon} + T_{\mu\lambda}(p_1,p_2,k)\right\} u_A\right], \tag{1}$$

where $D_{\mu\nu}$ is the photon propagator and $O_\lambda = \gamma_\lambda(1+\gamma_5)$. In writing this equation we have assumed that A is a negatively charged hyperon and B is neutral. If A is neutral and B is positively charged then one should replace $p_1 \to -p_2$ and change the overall sign in Eq. (1). For other combinations of charges of A and B appropiate changes must be done in Eq. (1).

Graph 1(d) contains the electron wave function renormalization. After mass renormalization it is given by

$$M_2 = \frac{\alpha}{8\pi^3 i}\frac{G_V}{\sqrt{2}}\bar{u}_B W_\lambda u_A \int d^4k D_{\mu\nu}(k)\bar{u}_\ell \frac{(2\ell_\mu - \gamma_\mu \not{k})\not{\ell}(2\ell_\nu - \not{k}\gamma_\nu)}{2m(k^2-2\ell\cdot k+i\varepsilon)^2}(\not{\ell}+m)O_\lambda v_\nu \tag{2}$$

Graph 1(e) can be written in general as

$$M_3 = \frac{\alpha}{2\pi} \bar{u}_\ell O_\lambda v_\nu \bar{u}_B \left[V_\lambda + A_\lambda\right] u_A , \tag{3}$$

where V_λ and A_λ can be expanded into 6 new form factors in exactly the same form as W_λ was expanded in Eq. (2.12). The infra-red divergence in M_3 can be extracted in a model independent way and is contained in

$$M_3^c = \frac{\alpha}{8\pi^3 i} \frac{G_v}{\sqrt{2}} \bar{u}_B W_\lambda u_A \bar{u}_\ell O_\lambda v_\nu \int d^4k\, D_{\mu\nu}(k) \frac{(2p_1-k)_\mu (2p_1-k)_\nu}{(k^2-2p_1\cdot k+i\varepsilon)^2} \tag{4}$$

as was shown by Meister and Yennie[3].

From Eq. (1) we can extract a model independent contribution which is finite in the ultraviolet and contains the infra-red divergence of graphs 1(a)-1(c); namely,

$$M_1^i = -\frac{G_v}{\sqrt{2}} \frac{\alpha}{4\pi^3 i} \int d^4k\, D_{\mu\nu}(k) \bar{u}_\ell \frac{2\ell_\nu - \gamma_\nu \not{k}}{k^2-2\ell\cdot k+i\varepsilon} O_\lambda v_\nu \quad \bar{u}_B \frac{W_\lambda (2p_{1\mu}-k_\mu)}{k^2-2p_1\cdot k+i\varepsilon} u_A \tag{5}$$

All the complications of virtual radiative corrections are contained in the second term of M_1 and in the remaining part of M_3, once M_3^c is extracted. The sum of M_1^i, M_2 and M_3^c is finite, model independent, it contains the infra-red divergence, and it is gauge invariant. The weak transition amplitude with virtual radiative corrections can then be expressed as

$$A_v = M_0 + M_1^i + M_2 + M_3^c + S, \tag{6}$$

where S is the model dependent part coming from the two terms just mentioned.

The inner-bremsstrahlung graphs can be treated in exact parallelism with the virtual ones. Their amplitudes are given by

$$M_B = \frac{eG_v}{\sqrt{2}} \bar{u}_B W_\mu u_A \bar{u}_\ell O_\mu v_\nu \left(\frac{2\ell\cdot\varepsilon}{2\ell\cdot k+\lambda^2+i\varepsilon} + \frac{2\varepsilon\cdot p_1 - \varepsilon\cdot k}{\lambda^2-2p_1\cdot k+i\varepsilon} \right) +$$

$$+ \frac{eG_v}{\sqrt{2}} \bar{u}_B W_\lambda u_A \bar{u}_\ell \frac{\not{\varepsilon}\not{k}}{2\ell\cdot k+i\varepsilon} O_\lambda v_\nu + \frac{eG_v}{\sqrt{2}} \bar{u}_B \varepsilon_\mu T_{\mu\lambda} u_A \bar{u}_\ell O_\lambda v_\nu . \tag{7}$$

ε_μ is the photon polarization four-vector and λ is a small photon mass. The first three terms are model independent and contain the infra-red divergence. The fourth one is model dependent. The definition of $T_{\mu\nu}$ is exactly as in Eq. (1), except that now the k that appears in it corresponds to a real photon and is subject to overall energy-momentum conservation. This $T_{\mu\nu}$ comes from graphs 2(a)-2(c), except that the second term in Eq. (7) has been extracted from graph 2(b). λ serves as an infra-red cut-off, and it will cancel its counterpart in Eq. (6).

We have now separated both the virtual and bremsstrahlung corrections into two parts: a model independent one —the sum of M_1^i, M_2, and M_3^c for the virtual corrections and the sum of the first three terms of Eq. (7) for the bremsstrahlung corrections— that can be computed and we shall do so in Sec. 4, and a model dependent part —S in Eq. (6) and the last term of Eq. (7)— that will be discussed in the next section.

5.3. Effective Form Factors.

We shall now discuss the model dependent part of the virtual corrections, given in S of Eq. (6). As we stressed in Sec. 1, we do not want to involve ourselves with a particular model of strong and weak interactions. Instead, we want to exploit general properties of $T_{\mu\lambda}$ as far as we can.

In Eq. (1) we can always write the lepton covariant in the form $\bar{u}_e O_\lambda v_\nu$ by relabeling indeces. After performing the integration over k we must end up with a one-index tensor, which we shall call simply T'_λ . The model dependent part of M_3 in Eq. (3) is already a one-index tensor and it can be included into S. Dropping the prime on T'_λ to indicate this inclusion, S becomes

$$S = \frac{\alpha}{\pi} \; \bar{u}_B T_\lambda (p_1, p_2, \ell) u_A \bar{u}_\ell O_\lambda v_\nu . \tag{8}$$

Using Lorentz covariance T_λ can be expanded in a general way into

$$\bar{u}_B T_\lambda u_A = \bar{u}_B \Big[c\gamma_\lambda + \frac{a}{M_1}\sigma_{\lambda\mu}q_\mu + \frac{b}{M_1}\sigma_{\lambda\mu}\ell_\mu + \frac{e}{M_1}\, q_\lambda + \frac{f}{M_1}\, \ell_\lambda + $$

$$+ \left(d\gamma_\lambda + \frac{a'}{M_1}\,\sigma_{\lambda\mu}q_\mu + \frac{b'}{M_1}\sigma_{\lambda\mu}\ell_\mu + \frac{e'}{M_1} q_\lambda + \frac{f'}{M_1}\,\ell_\lambda\right) \gamma_5 \Big] u_A + \dots \tag{9}$$

$c, a, \ldots, d, a', \ldots, f'$ are new form factors, functions of the invariant variables q^2, $\ell \cdot (p_1+p_2)$, and $\ell \cdot q$. The dots in Eq. (9) stand for the covariants which have ℓ_μ and q_μ simultaneously. We do not need their explicit form for our argument and we shall not give it. We have normalized the new form factors by M_1. Since the lepton mass can only appear inside a logarithm as was shown by Sirlin[1], we must not use it to normalize the form factors, otherwise non-logarithmic lepton mass singularities would be introduced. The normalization we use is supported by the analyticity of strong interactions, which we assume to be valid when strong and electromagnetic interactions interfere with one another. Since ℓ_μ is always of the order of q_μ, because of energy-momentum conservation, all the contributions that contain it explicitly in Eq. (9) can be ignored in the approximation in which terms of order $(\alpha/\pi)q$ can be neglected. Therefore, within this approximation, S takes the simple form.

$$S = \frac{\alpha}{\pi} \bar{u}_B (c\gamma_\lambda + d\gamma_\lambda\gamma_5) u_A \bar{u}_\ell O_\lambda v_\nu, \tag{10}$$

where c and d are constants.

Returning to Eq. (6), we see that c and d can be absorbed into M_o by redefining the leading form factors f_1 and g_1 as

$$f_1'(0) = f_1(0) + \frac{\alpha}{\pi} c, \tag{11}$$

and

$$g_1'(0) = g_1(0) + \frac{\alpha}{\pi} d \tag{12}$$

Eq. (6) can be rewritten as

$$A_V = M_o' + M_1^i + M_2 + M_3^c, \tag{13}$$

The prime on M_o' indicates that it is the effective form factors defined in Eqs. (11) and (12) that appear in it now.

c and d are finite within the framework of gauge theories. Otherwise they depend on the ultraviolet cut-off. We shall accept that they can be manipulated as ordinary algebraic quantities, as gauge theories assure us that we can do. The proof that this is the case has been done[4] assuming the validity of current algebra to handle the effects of strong interactions upon the vector part of the weak interaction vertex. In this respect we hope that strong interactions are in general well enough behaved as not to foul one of the best contributions of gauge theories into the radiative correction problem.

The model dependent part of the bremsstrahlung correction comes only from the term with $T_{\mu\lambda}$ in Eq. (7). It can be expanded into form factors as we have just done for S. The important point to notice now is that $T_{\mu\lambda}$ does not contain any infra-red divergence and that even more, because of gauge invariance, it is transverse, i.e., $k_\lambda T_{\mu\lambda} = 0$. Its Lorentz covariant expansion is of the form

$$\varepsilon_\mu T_{\mu\lambda}(p_1,p_2,k) = \frac{1}{-2p_1\cdot k+i}\left\{ 2\varepsilon\cdot p_1(w_1 p_+\cdot k + w_1' q\cdot k)\gamma_\lambda + \ldots\right\}$$

$$+ \frac{1}{2k\cdot p_2+i\varepsilon}\left\{ \varepsilon\cdot p_+ w_2 p_{+\lambda} + \ldots \right\}, \tag{14}$$

where $p_+ = p_1 + p_2$ and the dots stand for several dozens of terms. Again, we do not have to see their explicit form for our argument. What matters is that the two denominators that appear in Eq. (14) contain p_1 or p_2. This should be compared to the other terms in M_B, Eq. (7); it is ℓ that appears in the denominator of the other terms Now in the square of M_B when the form factors are properly normalized, as was done in the virtual model dependent part, we see that the contributions due to $T_{\mu\lambda}$ will always be proportional to k_{max}/M_1, while whatever multiplies this factor should be of the same order of magnitude as the terms that come from the model independent part of M_B. In other words, the contributions of $T_{\mu\lambda}$ to M_B are always suppressed by a factor k_{max}/M_1. Within our approximation of neglecting terms of order $(\alpha/\pi)q$, we can always neglect the contributions from $T_{\mu\lambda}$ to the square of M_B because k_{max} is of order q. Therefore, there is no model dependence contributed by M_B to $d\omega$. All the model dependence of radiative corrections is contained in f_1' and g_1' of Eqs. (11) and (12).

We are paying a price for not knowing how to compute the model dependent part of radiative corrections. But in a certain way it is a fair price. It is the effective form factors only that can be obtained in studying HSD and not the original ones we intended to get. The change comes mainly from what we do not know of strong interactions and of the details of weak interactions. It is therefore fair enough to study simultaneously f_1, g_1, c, and d through the effective form factors f_1' and g_1'.

5.4. Radiative Corrections to the Differential Decay Rate.

We are now in a position to calculate the radiative corrections to the differential decay of rate of HSD. We must calculate the Feynman integrals hidden in Eq. (13), square A_V, keeping only terms up to first order in α, and compute the resulting traces. Next, we must square the model independent part of M_B, Eq. (7), sum over the photon polarization ε_μ, again compute the resulting traces, and finally add it to the previous one. It sounds and it is easy, but it is very long and tedious. We shall not repeat the calculation here and shall limit ourselves to reproduce the final results.

When the initial hyperon is polarized the result is

$$d\omega(A\to B\ell\nu) = \frac{G_V^2}{(2\pi)^5}\,\frac{E\ell(E_m-E)^2 dE d\Omega_\ell d\Omega_\nu}{\left(1-\frac{E}{M_1}+\frac{\ell}{M_1}\,\hat{\ell}\cdot\hat{p}_\nu\right)^3}$$

$$\times\left\{D_1'\left[1+\frac{\alpha}{\pi}(\phi_1+\theta_1)\right]+\beta\hat{\ell}\cdot\hat{p}_\nu D_2'\left[1+\frac{\alpha}{\pi}(\phi_2+\theta_2)\right]+\right.$$

$$\left.+\beta\hat{s}_1\cdot\hat{\ell}D_3'\left[1+\frac{\alpha}{\pi}(\phi_2+\theta_2)\right]+\hat{s}_1\cdot\hat{p}_\nu D_4'\left[1+\frac{\alpha}{\pi}(\phi_1+\theta_1)\right]\right\} \qquad (15)$$

D_1', D_2', D_3', and D_4' are exactly the same expressions as in Eqs. (2.17)-(2.20), except that it is now f_1' and g_1' that appear in them instead of f_1 and g_1. The primes on these new D_i' (i = 1,2,3,4) are a reminder of this substitution. The model independent part of radiative corrections is contained in $\phi_1+\theta_1$ and $\phi_2+\theta_2$, whose explicit form are

$$\phi_1+\theta_1 = 2\left(\frac{1}{\beta}\tanh^{-1}\beta-1\right)\left[\frac{E_m-E}{3E}-\frac{3}{2}+\ln\frac{2(E_m-E)}{m}\right]+\frac{2}{\beta}L\left(\frac{2\beta}{1+\beta}\right)+$$

$$+\frac{1}{2\beta}\tanh^{-1}\beta\left[2(1+\beta^2)+\frac{(E_m-E)^2}{6E^2}-4\tanh^{-1}\beta\right]-$$

$$- \frac{3}{8} + \begin{cases} \frac{\pi^2}{\beta} + \frac{3}{2}\ell n\left[\frac{M_2}{m}\right] & \text{(n.d.h.)} \\ \frac{3}{2}\ell n\left(\frac{M_1}{m}\right) & \text{(c.d.h.)} \end{cases} \qquad (16)$$

and

$$\phi_2+\theta_2 = \left(\frac{1}{\beta}\tanh^{-1}\beta - 1\right)\left[\frac{(E_m-E)^2}{12\beta^2E^2} + \frac{2(E_m-E)}{3\beta^2E} + 2\ell n\,\frac{2(E_m-E)}{m} - 3\right] +$$

$$+ \frac{2}{\beta}L\left(\frac{2\beta}{1+\beta}\right) - \frac{2}{\beta}\tanh^{-1}\beta(\tanh^{-1}\beta - 1) -$$

$$- \frac{3}{8} + \begin{cases} \frac{\pi^2}{\beta} + \frac{3}{2}\ell n\left(\frac{M_2}{m}\right) & \text{(n.d.h.)} \\ \frac{3}{2}\ell n\left(\frac{M_1}{m}\right) & \text{(c.d.h.)} \end{cases} \qquad (17)$$

In these expressions β is the electron velocity, L is the Spence function, and E_m is the electron maximum energy. Everything is given in the rest frame of the initial hyperon. The abbreviation n.d.h. and c.d.h. stand for "neutral decaying hyperon" and "charged decaying hyperon", respectively. ϕ_1 and ϕ_2 come from the virtual radiative correction, while θ_1 and θ_2 come from the bremsstrahlung contribution. Eqs. (16) and (17) are free of the infra-red divergence cut-off λ, as was expected.

When the polarization of the emitted hyperon is observed, the differential decay rate, in the center of mass of B, is

$$d\omega(A \to B\ell\nu) = \frac{1}{2}\left(\frac{M_2}{M_1}\right)^3 \frac{G_V^2}{(2\pi)^5} \frac{(\hat{E}_m-\hat{E})^2\hat{\ell}\hat{E}d\hat{E}d\hat{\Omega}_\ell d\hat{\Omega}_\nu}{\left[1+\frac{\hat{E}}{M_2}(1-\beta\hat{x})\right]^3}$$

$$\times\left\{\hat{D}_1'\left[1+\frac{\alpha}{\pi}(\hat{\phi}_1+\hat{\theta}_1)\right] + \beta\hat{\ell}^*\cdot\hat{p}_\nu^*D_2'\left[1+\frac{\alpha}{\pi}(\hat{\phi}_2+\hat{\theta}_2)\right] + \right.$$

$$\left. + \beta\hat{s}_2\cdot\hat{\ell}^*\hat{D}_5'\left[1+\frac{\alpha}{\pi}(\hat{\phi}_2+\hat{\theta}_2)\right] + \hat{s}_2\cdot\hat{p}_\nu^*\hat{D}_6'\left[1+\frac{\alpha}{\pi}(\hat{\phi}_1+\hat{\theta}_1)\right]\right\} \qquad (18)$$

D_5' and D_6' are exactly the same expressions as in Eqs. (2.25) and (2.26), except that the replacement of f_1 and g_1 by f_1' and g_1' has been performed also, and again the primes on them should serve as a reminder. Other definitions are as before, but they apply to the center of mass of B. The model independent radiative corrections are

$$\hat{\phi}_1+\hat{\theta}_1 = 2\left(\frac{1}{\beta}\tanh^{-1}\beta-1\right)\left[\frac{\hat{E}_m-\hat{E}}{3\hat{E}}-\frac{3}{2}+\ell n\,\frac{2(\hat{E}_m-\hat{E})}{m}\right]+\frac{2}{\beta}L\left(\frac{2\beta}{1+\beta}\right)+$$

$$+\frac{1}{2\beta}\tanh^{-1}\beta\left[2(1+\beta^2)+\frac{(\hat{E}_m-\hat{E})^2}{6\hat{E}^2}-4\tanh^{-1}\beta\right]-\frac{3}{8}+$$

$$+\begin{cases}\frac{3}{2}\ell n\left(\frac{M_1}{m}\right) & \text{n.e.h.}\\[2ex] \frac{\pi^2}{\beta}+\frac{3}{2}\ell n\left(\frac{M_2}{m}\right) & \text{c.e.h.}\end{cases} \tag{19}$$

and

$$\phi_2+\theta_2 = \left(\frac{1}{\beta}\tanh^{-1}\beta-1\right)\left[\frac{(\hat{E}_m-\hat{E})^2}{12\beta^2\hat{E}^2}+\frac{2(\hat{E}_m-\hat{E})}{3\beta^2\hat{E}}+2\ell n\,\frac{2(\hat{E}_m-\hat{E})}{m}-3\right]+$$

$$+\frac{2}{\beta}L\left(\frac{2\beta}{1+\beta}\right)-\frac{2}{\beta}\tanh^{-1}\beta(\tanh^{-1}\beta-1)-\frac{3}{8}+$$

$$+\begin{cases}\frac{3}{2}\ell n\left(\frac{M_1}{m}\right) & \text{n.e.h.}\\[2ex] \frac{\pi^2}{\beta}+\frac{3}{2}\ell n\left(\frac{M_2}{m}\right) & \text{c.e.h.}\end{cases} \tag{20}$$

They formally agree with Eqs. (16) and (17), except that it is the variables with a "hat" that appear now. As a matter of fact, Eqs. (19) and (20) agree numerically with Eqs. (16) and (17), respectively, owing to our approximations. We could have simply limited ourselves to state this last fact, but for the sake of clarity it is better to look for oneself into Eqs. (19) and (20). n.e.h. and c.e.h mean "neutral emitted hyperon" and "charged emitted hyperon" respectively.

Looking at the long, long expressions of Sec. (2.3) and at the long expressions we have obtained in this section one can no less than be pleasantly surprised by the compactness of Eqs. (15) and (18). They are after all very simple and easy to read.

It is worthwhile emphasizing that it is the primed form factors that always appear in Eqs. (15) and (18). This means that it is only them that are experimentally accesible and not f_1 or g_1. In other words, as far as a form factors can be determined experimentally then it is f_1' and g_1' that can be determined in a model independent fashion. It should be clear that the model dependence of radiative corrections affects only theoretical physics and that, at least as experimental analysis is concerned, the vicious circle we mentioned in Sec. 1 is not a problem. In this sense we can claim that it has been opened. The model independent experimental values of f_1' and g_1' can be used to compare with theoretical predictions for f_1 and g_1 along with some model for c and d. If one chooses to use a different model to compute c and d one does not have to change the experimental values of f_1' and g_1'.

We shall say something about c and d in Sec. 6. Some reasonable estimates can be obtained within acceptable assumptions.

5.5. A Theorem for Integrated Observables.

In order to obtain the radiative corrections to the integrated observables we must integrate Eqs. (15) and (18) over the kinematical variables. From Eq. (15) we obtain[5,2.7] readily expressions for the corrected decay rate, $\alpha_{e\nu}$, α_e, and α_ν angular coefficients,

$$R = R^{0} \left(1 + \frac{\alpha}{\pi} \frac{\Phi_1}{b} \right) , \tag{21}$$

$$\alpha_{e\nu} = \alpha_{e\nu}^{0} \left[1 + \frac{\alpha}{\pi} \left(\frac{\Phi_2}{a} - \frac{\Phi_1}{b} \right) \right] , \tag{22}$$

$$\alpha_e = \alpha_e^0 \left[1 + \frac{\alpha}{\pi} \left(\frac{\Phi_2}{a} - \frac{\Phi_1}{b} \right) \right], \tag{23}$$

$$\alpha_\nu = \alpha_\nu^0, \tag{24}$$

R^0, $\alpha_{e\nu}^0$, α_e^0, and α_ν^0 are the uncorrected observables, as given in Chapter 3 and Appendix 1; except that f_1' and g_1' must appear in them, instead of f_1 and g_1. Φ_1 and Φ_2 come from the model independent part of radiative corrections. They are given, within our approximations, by

$$\Phi_1 = \int_m^{E_m} dE\, E\ell (E_m - E)^2 \, (\phi_1 + \theta_1), \tag{25}$$

$$\Phi_2 = \int_m^{E_m} dE\, \ell^2 (E_m - E)^2 \, (\phi_2 + \theta_2), \tag{26}$$

the above a and b are

$$a = \int_m^{E_m} dE\, \ell^2 (E_m - E)^2, \tag{27}$$

$$b = \int_m^{E_m} dE\, E\ell (E_m - E)^2, \tag{28}$$

In Table 1 we have tabulated the numerical values of $(\alpha/\pi)\Phi_1/b$ and $(\alpha/\pi)\Phi_2/a$, for electron-mode decays. In Table 2 we have done the same for some of the muon mode decays.

TABLE 1. Numerical values of the integrated model independent radiative corrections to the electron-mode semileptonic decays of hyperons. The percentage values are obtained multiplying by 100.

	$\frac{\alpha}{\pi}\Phi_1/b$	$\frac{\alpha}{\pi}\Phi_2/a$
np	0.0486	0.0474
Λp	0.0207	0.0196
$\Xi^0\Sigma^+$	0.0226	0.0215
$\Sigma^0 p$	0.0196	0.0184
$\Sigma^+\Lambda$	0.0015	0.0004
$\Sigma^-\Lambda$	0.0012	0.0001
$\Sigma^- n$	-0.0025	-0.0037
$\Xi^-\Lambda$	-0.0015	-0.0027
$\Xi^-\Sigma^0$	-0.0000	-0.0011
$\Xi^-\Xi^0$	0.0104	0.0100

TABLE 2. Numerical values of the integrated model independent radiative corrections to the muon-mode semileptonic decays of hyperons. The percentage values are obtained multiplying by 100.

	$\frac{\alpha}{\pi}\Phi_1/b$	$\frac{\alpha}{\pi}\Phi_2/a$
Λp	0.0468	0.0429
$\Sigma^0 p$	0.0336	0.0318
$\Sigma^- n$	-0.0022	-0.0011
$\Xi^-\Lambda$	-0.0013	0.0001
$\Xi^-\Sigma^0$	0.0002	0.0022

In the case of a polarized emitted hyperon, the radiative corrections to the electron and neutrino asymmetries are exactly as in Eqs. (23) and (24); namely,

$$\hat{\alpha}_\ell = \hat{\alpha}^o_\ell \left\{ 1 + \frac{\alpha}{\pi} \left[\frac{\hat{\Phi}_2}{\hat{a}} - \frac{\hat{\Phi}_1}{\hat{b}} \right] \right\} , \tag{29}$$

$$\hat{\alpha}_\nu = \hat{\alpha}^o_\nu , \tag{30}$$

The "hat" is a reminder that we are now in the center of mass of B. $\hat{a}$, $\hat{b}$, $\hat{\Phi}_1$, and $\hat{\Phi}_2$ are defined exactly as in Eqs. (25-28) by just putting "hats" all over

To get the radiative corrections to A and B requires a little more effort. The orthonormal basis to define them is

$$\hat{\alpha} = (\hat{\ell}^* + \hat{p}^*_\nu)/a' ,$$
$$\hat{\beta} = (\hat{\ell}^* - \hat{p}^*_\nu)/b' ,$$
$$\hat{n} = \hat{\beta} \times \hat{\alpha} ,$$

as we mentioned in Sec. (3.1). In order to use Eq. (18) to compute the radiative corrections to A and B, we change the angular variables of the electron to those of the vector $\hat{\alpha}$ using,

$$\hat{\ell}^* = -\hat{p}^*_\nu + (2\hat{p}^*_\nu \cdot \hat{\alpha})\hat{\alpha} .$$

The solid angle $d\hat{\Omega}_e$ is replaced by $(4\hat{p}^*_\nu \cdot \hat{\alpha}) d\Omega_\alpha$. The part containing $\hat{s}_2$ of Eq. (18) becomes, then,

$$d\omega(A \to B\ell\nu)_{\hat{s}_2} = \frac{1}{2} \left(\frac{M_2}{M_1} \right)^3 \frac{G_V^2}{(2\pi)^5} \frac{(\hat{E}_m - \hat{E})^2 \hat{\ell}\hat{E} d\hat{E} d\Omega_\alpha d\hat{\Omega}_\nu}{\left[1 + \frac{\hat{E}}{M_2} (1 - \beta\hat{x}) \right]^3}$$

$$\times (4\hat{p}^*_\nu \cdot \hat{\alpha}) \left\{ (2\hat{p}^*_\nu \cdot \hat{\alpha}) \hat{s}_2 \cdot \hat{\alpha} D'_5 + \hat{s}_2 \cdot \hat{p}^*_\nu (-D'_5 + D'_6) \right\} \tag{31}$$

Integrating Eq. (31) to get A, we obtain

$$R \times \hat{\alpha}_\alpha = \frac{G_V^2}{2\pi^3}\ \hat{a}\ \frac{2}{3}\left\{\bar{\hat{D}}'_5\left(1+\frac{\alpha}{\pi}\ \frac{\hat{\Phi}_2}{\hat{a}}\right)+\frac{\hat{b}}{\hat{a}}\bar{\hat{D}}'_6\left(1+\frac{\alpha}{\pi}\ \frac{\hat{\Phi}_1}{\hat{b}}\right)\right\}. \qquad (32)$$

where $\bar{\hat{D}}'_5$ and $\bar{\hat{D}}'_6$ are the integrated coefficients $\hat{D}'_5$ and $\hat{D}'_6$.

The radiative corrections to B are obtained by using

$$\hat{\ell}^* = \hat{p}^*_\nu - (2\hat{p}^*_\nu\cdot\hat{\beta})\hat{\beta}$$

to replace $\hat{\ell}^*$ and $d\hat{\Omega}_e$ in Eq. (18), so that $\hat{\beta}$ and $d\Omega_\beta$ appear instead. The result, after integrations, is

$$R \times \hat{\alpha}_\beta = \frac{G_V^2}{2\pi^3}\ \hat{a}\ \frac{2}{3}\left\{\bar{\hat{D}}'_5\left(1+\frac{\alpha}{\pi}\ \frac{\hat{\Phi}_2}{\hat{a}}\right)-\frac{\hat{b}}{\hat{a}}\bar{\hat{D}}'_6\left(1+\frac{\alpha}{\pi}\ \frac{\hat{\Phi}_1}{\hat{b}}\right)\right\}. \qquad (33)$$

Looking at the above results it turns out that a theorem can be proved. As a matter of fact, the V-A theory form of the integrated observables is not affected by radiative correction. In order to prove this theorem, we point out that when the electron mass can be neglected, it happens that

$$\phi_2 + \theta_2 - \phi_1 - \phi_1 = -\frac{(E_m - E)^2}{12E^2} \qquad (34)$$

then, the result of the integrations performed above becomes

$$\frac{\alpha}{\pi}\left(\frac{\Phi_2}{a} - \frac{\Phi_1}{b}\right) = -0.0012 \qquad (35)$$

When m cannot be neglected as is the case in the μ-mode decays and in the phase space integration of some e-mode cases, we get Table 3

TABLE 3. The result of the numerical integration for the model-independent radiative correction of Eqs. (25)-(28). The np and $\Xi^-\Xi^0$ stand for $n \to pe\nu$ and $\Xi^- \to \Xi^0 e\nu$; all others for muon modes, e.g., $\Lambda \to p\mu\nu$. The percentage values are obtained by multiplying by 100.

	np	$\Xi^-\Xi^0$	Λp	$\Sigma^0 p$	$\Sigma^- n$	$\Xi^-\Lambda$	$\Xi^-\Sigma^0$
$\frac{\alpha}{\pi}\left(\frac{\Phi_2}{a} - \frac{\Phi_1}{b}\right)$	0.0012	-0.0003	-0.0038	-0.0017	0.0011	0.0012	0.0020

Eq. (35) and Table 3 imply that within our approximations the difference between $(\alpha/\pi)\Phi_1/b$ and $(\alpha/\pi)\Phi_2/a$ (and between their "hatted" counter parts) can be neglected, i.e.,

$$\frac{\alpha}{\pi}\frac{\Phi_1}{b} \sim \frac{\alpha}{\pi}\frac{\Phi_2}{a}$$

Then Eqs. (22-24), (29-30) and (32-33) become

$$\begin{aligned} \alpha_{e\nu} &= \alpha^o_{e\nu} , \\ \alpha_e &= \alpha^o_e , \\ \alpha_\nu &= \alpha^o_\nu , \\ \hat{\alpha}_e &= \hat{\alpha}_e , \\ \hat{\alpha}_\nu &= \hat{\alpha}_\nu , \\ A &= A^o , \\ B &= B^o , \end{aligned} \tag{36}$$

both for e- and μ- mode cases. Eq. (21) for R remains, but since the radiative correction, is common to all terms in R_o, its V-A theory form is not changed either. The theorem is thus proved.

There remains only α_B. We shall not go into more details now but refer the interested reader to the literature[2.7]. It turns out that the bremsstrahlung part of Eq. (15) is not directly applicable to α_B. A numerical integration must be performed which is too lengthy to reproduce here. Suffice it to say that the numerical results obtained for α_B are also in agreement with the above theorem.

Let us stress that the model dependent corrections are still contained in Eqs. (21) and (36). The integrated observables are indeed affected by radiative correction, it is only their V-A form that is unchanged by them.

Before we close this section we would like to make some comments. We have strictly neglected all terms of order $(\alpha/\pi)q$. This can be relaxed if other assumptions are allowed. For example, the full phase space factors of Eqs. (15) and (18) can be kept in the numerical integrations of Eqs. (25)-(28). If this is done, our new assumptions could be that (i) the electron energy dependence of f_1' and g_1' is still negligible,

(ii) the model independent contributions to ϕ_i and θ_i are still well approximated by Eqs. (16), (17), (19), and (20), and (iii) the model dependent part of the inner-bremsstrahlung contribution is still negligible.

Of these three assumptions, the first two may be pretty good. It is only the last one that requires detailed study. It is conceivable that assumption (iii) is quite accurate. At any rate, an approach quite close to this one has been followed by Toth et al[9]. Numerical results have been reported. They agree within half of a percent with our estimates, and it seems that that difference is made even smaller when the full phase space factors are kept in Eqs. (25)-(28). It would be worthwhile to pursue the above assumptions further. Then, the expressions obtained would be useful for experiments with very high statistics —with millions of events.

5.6. The Model Dependent Part.

It is true that we cannot rigorously compute the model dependent part of radiative corrections. We can take two attitudes in this respect. One of them is to try to determine it from HSD data. The other one is to try to get reasonable estimates, that we may expect are good enough for low statistics experiments. We shall leave the first point of view for the next chapter and we shall address ourselves to the second point of view in this section.

We must choose a model for the details of weak interactions and for strong interactions. In a very careful analysis, Sirlin[4], using the Weinberg-Salam $SU(2) \times U(1)$ model and a current algebra approach, has estimated the model dependent part c, that affects the vector-current matrix-element in $n \to pe\nu$, as

$$\frac{\alpha}{\pi} c = \frac{\alpha}{4\pi} \left\{ 3\ln\left(\frac{M_z}{M_p}\right) + 6\bar{Q}\ln\left(\frac{M_z}{M}\right) + 2C' + A\bar{g} \right\}. \tag{37}$$

The first term is a universal photonic contribution that arises from the V_μ current, the second and third terms correspond to photonic corrections induced by the A_μ current and $A\bar{g}$ is induced by strong interactions in the asymptotic domain. $\bar{Q}$ is the average u and d quark charges, M_z is the Z_o mass, M_p is the proton mass, and M is a a hadronic mass of the order of the A_1-resonance mass.

The estimates[6] for C' and $A\bar{g}$ make it plausible that their contributions be negligible, some 30 to 40 times smaller than other contributions in Eq. (37). Accepting these estimates we can then ignore these two terms, within our approximations. Eq. (37) becomes then, redefining c to absorb its coefficients,

$$C \equiv \frac{\alpha}{\pi} c \simeq \frac{\alpha}{4\pi} \left\{ 3\ell n \left(\frac{M_z}{M_p} \right) + 6\bar{Q}\ell n \left(\frac{M_z}{M} \right) \right\}. \tag{38}$$

The main contribution to C comes from gauge theories, and the only reminiscence of strong interactions is in M and in $\bar{Q}$. This approximation of C is due mainly to the "leading logarithmic terms of the Z_0 mass".

In the SU(2) × U(1) model, with a standard SU(3) for the color symmetry, $\bar{Q} = 1/6$. Using $M_z = 91.5$ from Eq. (2.10) which comes from assuming the Weinberg angle at $\sin^2\theta_W = 0.23$, Eq. (2.8), and letting $M_p \simeq M \simeq 1$ GeV, we get

$$C \simeq 0.0105 = 1.05\%. \tag{39}$$

This value of C could give a noticeable contribution in HSD, although it may not yet be seen owing to the low statistics of the available data.

The model dependent correction d to the axial-vector current matrix element can be estimated also. One can prove that Eq. (37) applies with almost as good accuracy to the A_μ matrix element as it does to the V_μ case. An explicit calculation[7] in the case of the pseudoscalar meson leptonic decays, which are mediated by A_μ currents only, shows that Eq. (37) is also obtained. The mass M now is of the order of the mass of a strong interaction vector meson, which is some 20% smaller than 1 GeV, and can thus be taken at 1 GeV in our approximations. The estimates of the last two terms of Eq. (37) are also expected to be very small, provided the A_μ current is partially conserved. Sirlin has stated[8] the above conclusion in a general fashion with a theorem that asserts that the behavior of Eq. (37) for large M_z and M_W masses is common to V_μ and A_μ currents and is given by Eq. (38), essentially. From this point of view we can conclude, then, that

$$\frac{\alpha}{\pi} d \simeq C,$$

or

$$d \simeq c. \tag{40}$$

The above estimates are not rigorous. One should really prove that the terms neglected are as small as they have been estimated. At any rate these estimates seem reasonable. We can at least use them to believe that the difference between c and d is small enough that we can equate them within our approximations. Also, the difference between c in one HSD and in another one, which would appear through M in Eq. (38), should be small enough as to be negligible. According to these estimates and, again, within our approximations, we can conclude that the main role of the model dependent part of radiative corrections is to affect the value of G_μ, which is common to all HSD.

It then seems reasonable to propose as a parametrization of the model dependent part of radiative corrections a modified Fermi coupling constant,

$$G_\mu \equiv G_\mu(1+C). \tag{41}$$

C could be left as a free parameter and one could attempt to obtain it from HSD data. The value obtained should then be compared with the estimate of Eq. (39).

Chapter 6. Other Corrections

6.1. Momentum-Transfer and Form Factors.

In addition to ignoring radiative corrections, another assumption in the comparison of Sec. 4.4 of CT and HSD data was that the q^2-dependence of the several form factors involved is not important. This is quite to be expected for the induced form factors, but not so for the leading ones.

As we discussed in Sec. 3.1, the vector and axial-vector leading from factors can be expanded to first order in q^2. Higher powers of q^2 would not be required. The two parameters λ_1^f and λ_1^g are introduced this way —see Eq. (3.11) and (3.12). Let us first discuss λ_1^f, which we shall often refer to as the "V-slope".

If the SU_3-octet assumption for the V_μ current is valid, then Eq. (3.11) has explicit SU(3) indeces and Clebsch-Gordan coefficients,

$$f_1^{AB}(q^2) = C_F^{AB} F_1(q^2) + C_D^{AB} D_1(q^2) =$$

$$= C_F^{AB}\left[F_1(0) + \lambda_{F_1} q^2\right] + C_D^{AB}\left[D_1(0) + \lambda_{D_1} q^2\right], \tag{1}$$

in accordance with Eq. (4.4). In Eq. (1) we have expanded the reduced form factors to first order in q^2. We see then that, if the symmetry limit is valid, there are two V-slopes for all the HSD: λ_{F_1} and λ_{D_1}.

The connection with the electromagnetic current through the CVC hypothesis, along with the validity of the symmetry limit, determines λ_{F_1} in terms of the measured q^2-dependence of the charge form factors of the proton and the neutron[3.12]. The result is

$$\lambda_{F_1} = 6.13\ \text{GeV}^{-2},$$

and

$$\lambda_{D_1} = 0.12\ \text{GeV}^{-2}. \tag{2}$$

These two values, Eqs. (1), and (3.11) determine λ_1^f, for each HSD.

There is an ambiguity in the above procedure. λ_1^f is dimensionless, while λ_{F_1} and

λ_{G_1} are dimensioned. The question is which mass M should we use to take care of the dimensions of q^2 in $\lambda_1^f q^2/M^2$. In Eq. (3.11) we agreed to put M_1 in place of M. If we are consistent with our discussion of the determination of f_2 in Sec. (4.3) through CVC, then, we should use the mass of nucleons instead of M_1. This is what we shall do for definiteness, although the ambiguity is not quite solved. The reason is that the experimental numbers for the proton and neutron electromagnetic form factors include already symmetry-breaking effects. Therefore, in order to use SU(3) symmetry, one should first correct those numbers and extract their symmetry limit values, and then perform the necessary rotations to get the symmetric values of λ_{F_1} and λ_{D_1}. But since the contributions of the V-slopes to the observables amount to a few percent anyway, the bias introduced in them is very small (around 0.2 of a few percent). We can take Eqs. (2) as good estimates for our purposes and leave further refinements for a future more precise experimental situation.

An equation analogous to Eq. (1) applies to g_1; namely,

$$g_1^{AB}(q^2) = C_F^{AB}\left[F(0) + q^2\lambda_F\right] + C_D^{AB}\left[D(0) + q^2\lambda_D\right] . \tag{3}$$

There are two "A-slopes" for all the leading axial-vector form factors of HSD. Unfortunately, no equivalence of CVC exists for the A_μ current, nor enough experimental evidence is available on the q^2-dependence of g_1 for several reactions or decays. Only for the reaction $\nu_\mu + p \to n + \mu$ the q^2-dependence of g_1 has been measured with some precision[3.12]. From here one can extract the q^2-dependence of g_1 in $n \to pe\nu$, assuming e-μ universality as CT does. The following dipole parametrization has been used,

$$g_1^{np}(q^2) = \frac{g_1^{np}(0)}{(1-q^2/M_A^2)^2} . \tag{4}$$

The measured slope is the inverse of the square of a certain mass M_A. The most recent value for M_A is

$$M_A = 0.96 \pm 0.03 \text{ GeV}. \tag{5}$$

Expanding Eq. (4) to first order in q^2 and comparing with Eq. (3), we get one relation for the A-slopes; namely,

$$\lambda_D = \frac{1}{\sqrt{0.3}} \left[\frac{\lambda_F}{\sqrt{6}} - \frac{2 \times 1.26}{(0.96)^2} \right] . \qquad (6)$$

We have put $g_1(0) = 1.26$. In deriving this equation, we have assumed in analogy to the case of the V-slopes that symmetry-breaking effects can be ignored. Eq. (6) is good enough for our purposes.

There is a different approach to deal with the V- and A- slopes. It is motivated by analyticity of strong interactions. If a singularity such as a pole is lying near the region of q^2 which is of kinematical interest, then, the relevant q^2-dependence would be dominated by that singularity. This approximation is very often welcome by experimentalists. We shall discuss it in Sec. 3.

6.2. Partial Conservation of the Axial-Vector Current.

For μ-mode HSD there is one contribution that was not determined in Sec. 4.3, that of the induced pseudoscalar form factor g_3. CT does not say anything about it. It is non-zero, in principle, in contrast to f_3, which is ruled out by the absence of second-class currents and the validity of the symmetry limit postulates.

g_3 is determined if the A_μ current is partially conserved (PCAC). We shall not go into details here, but shall refer the reader to the literature through Ref. (3.12). PCAC connects g_3 to $g_1(q^2)$. The knowledge of the latter fixes g_3. Explicitly, we get from Ref. (3.12) that

$$g_3(0) = -2M_N g_1(0) \left[\frac{1}{m_i^2} - \frac{g_1'(0)}{g_1(0)} \right] . \qquad (7)$$

M_N is the nucleon mass, m_i is the pion or kaon mass if $\Delta s = 0$ or $|\Delta s| = 1$, respectively, and $g_1'(0)$ is the derivative of $g_1(q^2)$ at $q^2 = 0$.

The contribution of g_3 to μ-mode decays is potentially important because of the presence of m_i in the denominator of the first term of Eq. (7). Order of magnitude, g_3 is some 7 to 10 times larger than g_1 if $\Delta S = 0$, and some 3 to 4 times larger if

$|\Delta S| = 1$.

6.3. Undetermined Contributions and Ambiguities.

Let us summarize the consequences of the hidden or, as we called them in Sec. 4.2, working-assumptions in the CT. If the experimental data is precise enough —as seems to be the case from the results of Sec. 4.4— one must account in the CT for radiative corrections, the q^2-dependence of the leading form factors, and for the induced pseudo-scalar form factor g_3.

Unfortunately, as we have seen in this and the last chapter these contributions are not all well determined. The radiative corrections introduce model dependent corrections, which can be reduced (if some reasonable arguments are accepted) to one unknown constant C, defined in Sec. 5.6. This constant affects the overall Fermi coupling constant in a multiplicative fashion. The A-slopes cannot yet be fully determined; owing to the lack of detailed measurements of $g_1(q^2)$ in β decays, λ_F in Eq. (6) is left free. And, finally, an additional assumption, past the CT, must be introduced to fix g_3, namely, PCAC. Otherwise g_3 has at least two undetermined parameter in the CT: its corresponding reduced form factors F_3 and D_3.

The consequences of the working-assumptions are, therefore, that in addition to the unknown parameters F, D, and θ, one has C, λ_F, F_3, and D_3. That is, in practice the CT has 7 free parameters.

We have several choices in front of us. The best one is to try to determine all of them from the experimental data, if at all possible. The values obtained for them could be compared afterwards with the theoretical estimates available for them, e.g., compare whatever is obtained for C with the result in Eq. (5.39). Another option is to fix the last four parameters in terms of their theoretical estimates and then extract F, D, and θ from experiment. Both options can be confronted later on.

One interesting point is to analyze carefully the A-slopes. Motivated by the parametrization of the q^2-dependence of the electromagnetic form factors of the nucleons, the question arises if $g_1(q^2)$ follows a monopole behavior —as suggested by pole dominance— or if it obeys a dipole behavior. The parametrization of $g_1^{np}(q^2)$ in

Eq. (4) implies already a dipole behavior. Pole-dominance gives

$$g_1^{AB}(q^2) = g_1^{AB}(0)\left(1 + \frac{q^2}{M_A'^2}\right), \tag{8}$$

to first order in q^2. For $\Delta S = 0$ decays $M_A' = 0.96$ GeV is expected, and for $|\Delta S| = 1$ decays $M_A' = 1.11$ GeV is expected, allowing for some symmetry breaking. Eq. (8) fixes then λ_F. For a dipole approach, Eq. (4) becomes for other HSD

$$g_1^{AB}(q^2) = g_1^{AB}(0)\left(1 + \frac{2q^2}{M_A^2}\right) \tag{9}$$

this also fixes λ_F, but at a different value than pole-dominance. The V-slopes can also be fixed by this procedure. As we mentioned at the end of Sec. 1., $f_1(q^2)$ would be given by

$$f_1^{AB}(q^2) = f_1^{AB}(0)\left(1 + a\frac{q^2}{M_V^2}\right), \tag{10}$$

with $M_V = 0.84$ MeV for $\Delta S = 0$ decays and $M_V = 0.97$ MeV for $|\Delta S| = 1$ decays, and $a = 1$ for pole-dominance or $a = 2$ for dipole behavior. The above approach is very often used in experimental analysis [1.9]. Let us just mention that so far there is no theoretical reason for Eq. (9) —just as no convincing explanation of the dipole behavior of the nucleon electromagnetic form factors has yet been produced. At any rate, it is clear that the monopole-dipole choice amounts to an ambiguity.

On the other hand, the determination of F_3 and D_3 of g_3 is a well defined problem. The question here is if there is enough experimental evidence on μ-mode decays that may give an idea on what they really are. Their values would be useful to test the PCAC assumption. Keeping this and the above questions in mind we now return to the experimental evidence on HSD.

Chapter 7. Comparison between Theory and Experiment

7.1. Relevance of the Corrections.

The effects of the model independent part of radiative corrections —the Φ_1 of Tables 5.1 and 5.2— on the comparison[1] of CT with the experimental evidence on HSD can be seen in columns (a) of Table 1. χ^2 is reduced by 11.5 from its value of 44.3 in Table 4.2. The main change comes from $n \to pe\nu$. Earlier, its transition rate was showing some worrisome deviation from its theoretical prediction. Clearly, radiative corrections are important. Their incorporation is well detected by HSD data and the agreement of CT is sensibly improved, although not enough. The deviation in $\Sigma^- \to \Lambda e\nu$ rate is reduced a little bit only and it remains almost as before. We have ignored the model dependent part of radiative corrections for the moment. We shall return to it in Secs. 2 and 3. At any rate we can conclude that HSD cannot be studied properly if radiative corrections are neglected.

To appreciate the effect of the V-slopes alone, we have incorporated them while leaving still other corrections out. The result for pole-dominance is displayed in columns (b) of Table 1. Using the CVC predictions or the dipole behavior for the V-slopes, as discussed in Sections 6.1 and 6.3, gives practically the same result as the pole-dominance case. The V-slopes also work in the direction of improving the agreement between theory and experiment. χ^2 is reduced by 4.1. This is not much, though, but still it means that the V-slopes must be taken into account because their contribution is noticeable. It is interesting to remark that the values of F, D, and θ and Tables 4.2 and 1 change very little, especially the Cabibbo angle. We also reproduced the predictions for the g_1/f_1 ratios. One can see that they have remained almost insensitive to the V-slopes.

Concerning the effects of the A-slopes, we have performed three fits, excluding radiative corrections and V-slopes. We have followed the approach outlined at the end of Sec. 6.3. In the first fit we used λ_F as a free parameter, in the second one we have fixed the A-slopes as given by pole-dominance, and in the third one we put the A-slopes as given by the dipole approach. The corresponding χ^2 were 42.9, 44.3, and 44.1. In none of these cases is there a significant improvement. The predictions for

TABLE 1. (a) Effect of radiative corrections. The model-dependent part of radiative corrections has been neglected. There is no contribution of the V- and A-slopes. Other conventions are as in Table 4.2. (b) Effect of V-slopes. Radiative corrections and A-slopes are not accounted for. Other conventions are as in (a).

	(a)		(b)	
	Prediction	$\Delta\chi^2$	Prediction	$\Delta\chi^2$
$n \to pe\nu$ (rate)	1.078	0.05	1.045	7.56
$\Sigma^+ \to \Lambda e\nu$ (rate)	0.284	0.29	0.281	0.24
$\Sigma^- \to \Lambda e\nu$ (rate)	0.470	2.94	0.465	2.46
$\Lambda \to pe\nu$ (rate)	3.191	0.10	3.173	0.00
$\Sigma^- \to ne\nu$ (rate)	7.049	0.74	7.014	0.97
$\Xi^- \to \Lambda e\nu$ (rate)	2.803	2.38	2.911	2.70
$\Xi^- \to \Lambda,\Sigma^o e\nu$ (rate)	3.335	0.38	3.439	0.29
$\Lambda \to p\mu\nu$ (rate)	0.526	0.32	0.520	0.38
$\Sigma^- \to n\mu\nu$ (rate)	3.127	0.10	3.252	0.62
$\Xi^- \to \Lambda\mu\nu$ (rate)	0.759	0.41	0.828	0.37
$n \to pe\nu$ ($\alpha_{e\nu}$)	-0.076	0.17	-0.078	1.10
$n \to pe\nu$ (α_e)	-0.083	0.01	-0.086	2.87
$n \to pe\nu$ (α_ν)	0.989	0.14	0.988	0.17
$\Sigma^\pm \to \Lambda e\nu$ ($\alpha_{e\nu}$)	-0.400	0.11	-0.400	0.11
$\Sigma^- \to \Lambda e\nu$ ($\alpha_{e\nu}$)	-0.403	0.15	-0.403	0.15
$\Sigma^- \to ne\nu$ ($\alpha_{e\nu}$)	0.344	1.64	0.339	1.39
$\Sigma^- \to ne\nu$ (α_e)	-0.684	7.18	-0.635	6.25
$\Lambda \to pe\nu$ ($\alpha_{e\nu}$)	0.024	5.36	0.007	1.91
$\Lambda \to pe\nu$ (α_e)	0.021	2.49	0.021	2.51
$\Lambda \to pe\nu$ (α_ν)	0.975	6.62	0.976	6.69
$\Lambda \to pe\nu$ (α_p)	-0.580	1.23	-0.583	1.34
$(g_1/f_1)_{np}$	1.256*	0.01	1.267*	4.05
$(g_1/f_1)_{\Lambda p}$	0.706*	0.23	0.721*	0.84
$(g_1/f_1)_{\Sigma^- n}$	-0.392*	1.53	-0.370*	3.41
$(f_1/g_1)_{\Sigma^-\Lambda}$	0.000*	0.34	0.000*	0.34

$F_{(a)} = 1.058 \pm 0.019$ $D_{(a)} = -1.504 \pm 0.016$ $\sin\theta_{(a)} = 0.233 \pm 0.002$ $\chi^2_{(a)} = 32.83$

$F_{(b)} = 1.098 \pm 0.020$ $D_{(b)} = -1.495 \pm 0.017$ $\sin\theta_{(b)} = 0.230 \pm 0.002$ $\chi^2_{(b)} = 40.27$

the observables have remained as in Table 4.2. Because of this, we do not reproduce them in another table. Leaving λ_F free was the only case in which χ^2 was slightly reduced, but an unacceptable negative λg_1 slope for $\Xi^- \to \Lambda e\nu$ turned out. It is clear that HSD data do not seem to require the presence of the A-slopes alone. Also the diference between pole-dominance and dipole slopes does not seem to matter much yet.

We now come to the incorporation of g_3 to the μ-modes. There are two possibilities. We have left F_3 and D_3 free and we have used the PCAC prediction of Eq. (6.7). In neither of these two trials could we see any important difference. It turned out that fixing F_3 and D_3 by using the PCAC prediction for g_3 amounted to the same result as keeping g_3 out as before: no change at all obtained. Leaving F_3 and D_3 did show a very small effect, but at the expense of very large values of g_3, some ten times larger than the PCAC prediction for them. Therefore, we may conclude that the μ-mode data are still far from showing any sensitivity to g_3. The role of the μ-mode rate is then to just check on e-μ universality. In what follows are shall simply ignore g_3.

7.2. Combined Effect.

Let us now incorporate in the CT the radiative corrections, the V-slopes and the A-slopes simultaneously. The results are shown in columns (a) and (b) of Table 2. We can see that the A-slopes, which by themselves did not improve the agreement between theory and experiment, now do provide an improvement. This can best be appreciated in the $\Sigma^- \to \Lambda e\nu$ rate. In Table 4.2, its predicted value exceeds its measured value and the A-slope contribution to it, being of positive sign, tends to increase this deviation. But when radiative corrections and V-slopes are also incorporated the corresponding $\Delta\chi^2$ drops. With variable λ_F the improvement is even better but again λg_1 of $\Xi^- \to \Lambda e\nu$ turned out unacceptably negative. For this reason we have not displayed the fit with variable λ_F in Table 2. The reason for the improvement of the contribution of the A-slopes is that the three contributions together allow for some extra freedom for the CT parameters F, D, and θ. In particular, D is allowed to decrease. This decrease is apparently very small, but it is enough to compensate the A-slope in $\Sigma^- \to \Lambda e\nu$ and still give a net reduction of its transition rate. At any

TABLE 2. Combined effect of radiative corrections, V-, and A-slopes. (a) Corresponds to pole-dominace A-slopes, (b) corresponds to dipole A-slopes, and (c) corresponds to fitted model dependent radiative correction parameter with dipole A-slopes. Other conventions are as in Table 4.2.

	(a) Prediction	(a) $\Delta\chi^2$	(b) Prediction	(b) $\Delta\chi^2$	(c) Prediction	(c) $\Delta\chi^2$
$n \to pe\nu$ (rate)	1.078	0.05	1.078	0.07	1.077	0.08
$\Sigma^+ \to \Lambda e\nu$ (rate)	0.278	0.20	0.280	0.22	0.280	0.22
$\Sigma^- \to \Lambda e\nu$ (rate)	0.462	2.12	0.464	2.35	0.464	2.35
$\Lambda \to pe\nu$ (rate)	3.185	0.04	3.184	0.04	3.184	0.04
$\Sigma^- \to ne\nu$ (rate)	7.026	0.89	7.029	0.87	7.029	0.87
$\Xi^- \to \Lambda e\nu$ (rate)	2.866	2.67	2.841	2.55	2.841	2.55
$\Xi^- \to \Lambda,\Sigma^o e\nu$ (rate)	3.383	0.34	3.356	0.36	3.356	0.36
$\Lambda \to p\mu\nu$ (rate)	0.539	0.22	0.542	0.20	0.542	0.20
$\Sigma^- \to n\mu\nu$ (rate)	3.266	0.70	3.277	0.77	3.277	0.77
$\Xi^- \to \Lambda\mu\nu$ (rate)	0.816	0.38	0.810	0.38	0.810	0.38
$n \to pe\nu$ ($\alpha_{e\nu}$)	-0.075	0.10	-0.075	0.08	-0.075	0.08
$n \to pe\nu$ (α_e)	-0.083	0.01	-0.083	0.05	-0.083	0.05
$n \to pe\nu$ (α_ν)	0.989	0.14	0.989	0.14	0.989	0.14
$\Sigma^\pm \to \Lambda e\nu$ ($\alpha_{e\nu}$)	-0.403	0.12	-0.408	0.15	-0.408	0.15
$\Sigma^- \to \Lambda e\nu$ ($\alpha_{e\nu}$)	-0.407	0.18	-0.412	0.22	-0.412	0.22
$\Sigma^- \to ne\nu$ ($\alpha_{e\nu}$)	0.318	0.56	0.299	0.14	0.299	0.14
$\Sigma^- \to ne\nu$ (α_e)	-0.646	6.45	-0.654	6.60	-0.654	6.60
$\Lambda \to pe\nu$ ($\alpha_{e\nu}$)	0.000	1.02	-0.014	0.04	-0.014	0.04
$\Lambda \to pe\nu$ (α_e)	0.018	2.61	0.012	2.93	0.012	2.93
$\Lambda \to pe\nu$ (α_ν)	0.975	6.58	0.976	6.65	0.976	6.65
$\Lambda \to pe\nu$ (α_p)	-0.582	1.28	-0.579	1.18	-0.579	1.18
$(g_1/f_1)_{np}$	1.254*	0.03	1.253*	0.10	1.253*	0.10
$(g_1/f_1)_{\Lambda p}$	0.712*	0.41	0.711*	0.39	0.711*	0.39
$(g_1/f_1)_{\Sigma^- n}$	-0.373*	3.18	-0.372*	3.21	-0.372*	3.20
$(f_1/g_1)_{\Sigma^-\Lambda}$	0.000*	0.34	0.000*	0.34	0.000*	0.34

$F_{(a)} = 1.080$ $D_{(a)} = -1.485$ $\sin\theta_{(a)} = 0.229$ $\chi^2_{(a)} = 26.68$

$F_{(b)} = 1.079$ $D_{(b)} = -1.484$ $\sin\theta_{(b)} = 0.227$ $\chi^2_{(b)} = 26.00$

$F_{(c)} = 1.079$ $D_{(c)} = -1.484$ $\sin\theta_{(c)} = 0.227$ $\chi^2_{(c)} = 26.00$

$C = 0.000 \pm 0.007$

rate, the deviation in the $\Sigma^- \to \Lambda e\nu$ rate remains noticeable. The several choices for the V-slopes give no practical difference in the results. Therefore, we shall restrict ourselves to the dipole case for these slopes from here on. This is already the case in Table 2(c).

Going through Table 2 and comparing it with Table 4.2 we can find many other improvements all over except with the polarization data of $\Lambda \to pe\nu$ and $\Sigma^- \to ne\nu$. There has been no noticeable change, in any fit considered, of α_ν or α_e in $\Lambda \to pe\nu$. Also, α_e in $\Sigma^- \to ne\nu$ remains still too far from the current experimental value.

We can now return to the model dependent part of radiative corrections. In columns (c) of Table 2, we show the result of letting C, defined in Eq. (5.38), be a free parameter along with F, D, and θ. We have used the dipole form for the A-slopes. If pole-dominance A-slopes are used in column (c), the results remain practically unchanged. It is interesting to notice that a positive non-zero C —albeit small— is allowed within error bars. In this sense we can say that C could be measured with more precise and accurate HSD data.

In Table 2(c) we have also computed the error bars for the CT parameters. Clearly they are all very much constrained. The experimental evidence of Table (3.1) does put the CT to genuine testing. Evidently, the working assumptions to compare CT with HSD must be relaxed by adding the corresponding corrections appropiately. Once they are incorporated comparison with experiment becomes meaningful. In this respect, what we have done so far is to improved the theoretical predictions. We must still look more critically into the agreement of CT with experiment, mainly because what up to now is the best prediction of CT —Table 2(c)— is not yet satisfactory.

7.3. Partition of the Experimental Data.

We have two questions in mind. The first one is what pieces of data do not agree well with the CT predictions and the second one is to what extent are such predictions stable. The two questions are related between themselves. The predictive power of CT is limited because of the appearance of three free parameters, F, D, and θ. It is possible that if some piece of data is not to agree with CT —because it is either wrong or because it is to show a genuine discrepancy— then F, D, and θ might be forced into a strange trend, in order that a minimum of χ^2 be found, which reproduces that particular piece of data. This is the danger of χ^2-fitting. One may be totally misled.

Looking through Tables 1, 2, and 4.2, one remarks that there is a well separated subset of the data which carries most of the weight of the deviations from theory. This subset is mainly composed of the polarization asymmetries in $\Lambda \rightarrow pe\nu$ and $\Sigma^- \rightarrow ne\nu$. None of the corrections introduced before, nor their combined effect gave a substantial change in the theoretical predictions for them. Their contribution to χ^2 has remained almost unchanged, despite the fact that χ^2 has been reduced to 2/3 of its value in Table 4.2.

We shall now compare the CT, including all the corrections and a variable C to (i) the transition rates alone, (ii) the transition rates and the g_1/f_1 ratios, and (iii) the transition rates and the angular-correlation coefficients, excluding the polarization asymmetries. The results are shown in Table 3. They are very illustrative. When only rates are used important changes of the CT parameters are allowed and even an unacceptable large value for C is obtained. When rates and g_1/f_1 ratios are used C comes out even smaller than in Sec. 2 and the $\Sigma^- \rightarrow \Lambda e\nu$ rate shows once more some deviation. In contrast, when rates and angular-correlation coefficients are used C comes out in the direction of its gauge-theory prediction, and the $\Sigma^- \rightarrow \Lambda e\nu$ rate remains at the same agreement as before. We have reproduced in Table 3 the predictions for those quantities that were left out in each one of the fits along with their potential contribution to χ^2. We have used dipole V and A-slopes; except for the g_1/f_1 ratios, for which we did not attempt to make any corrections for slopes, because of the inconsistencies mentioned in the Introduction. If pole-dominance A-slopes are used, the same pattern is obtained. Fits with variable λ_F give negative A-slopes for $\Xi^- \rightarrow \Lambda e\nu$. There is no need to reproduce all these fits here.

Clearly, the experimental information on the transition rates alone is insufficient,

TABLE 3. Partition of HSD data. (a) corresponds to fitting transition rates only, (b) corresponds to fitting transition rates and g_1/f_1 ratios, and (c) corresponds to fitting transition rates and $\alpha_{e\nu}$ coefficients only. In all cases radiative corrections were included, the model-dependent parameter C was fitted, and dipole A-slopes were used. Other conventions are as in Table 4.2.

	(a) Prediction	(a) $\Delta\chi^2$	(b) Prediction	(b) $\Delta\chi^2$	(c) Prediction	(c) $\Delta\chi^2$
$n \to pe\nu$ (rate)	1.080	0.01	1.077	0.11	1.077	0.08
$\Sigma^+ \to \Lambda e\nu$ (rate)	0.262	0.04	0.285	0.30	0.280	0.22
Σ^- $\Lambda e\nu$ (rate)	0.436	0.49	0.472	3.15	0.464	2.37
$\Lambda \to pe\nu$ (rate)	3.172	0.00	3.162	0.03	3.181	0.02
$\Sigma^- \to ne\nu$ (rate)	7.086	0.53	7.194	0.12	7.052	0.72
$\Xi^- \to \Lambda e\nu$ (rate)	3.113	4.02	2.815	2.43	2.846	2.58
$\Xi^- \to \Lambda,\Sigma^o e\nu$ (rate)	3.599	0.17	3.331	0.39	3.360	0.40
$\Lambda \to p\mu\nu$ (rate)	0.540	0.21	0.538	0.23	0.541	0.20
$\Sigma^- \to n\mu\nu$ (rate)	3.311	1.00	3.352	1.33	3.287	0.84
$\Xi^- \to \Lambda\mu\nu$ (rate)	0.887	0.34	0.802	0.39	0.811	0.38
$n \to pe\nu$ ($\alpha_{e\nu}$)	-0.047*	46.93	-0.076*	0.15	-0.074	0.08
$n \to pe\nu$ (α_e)	-0.049*	292.90	-0.083*	0.00	-0.082*	0.49
$n \to pe\nu$ (α_ν)	0.996*	0.06	0.989*	0.14	0.989*	0.13
$\Sigma^\pm \to \Lambda e\nu$ ($\alpha_{e\nu}$)	-0.408*	0.15	-0.408*	0.15	-0.408	0.15
$\Sigma^- \to \Lambda e\nu$ ($\alpha_{e\nu}$)	-0.412*	0.23	-0.412*	0.22	-0.412	0.22
$\Sigma^- \to ne\nu$ ($\alpha_{e\nu}$)	0.379*	3.94	0.277*	0.03	0.298	0.13
$\Sigma^- \to ne\nu$ (α_e)	-0.566*	5.03	-0.675*	7.02	-0.654*	6.61
$\Lambda \to pe\nu$ ($\alpha_{e\nu}$)	0.037*	9.45	-0.011*	0.14	-0.012	0.10
$\Lambda \to pe\nu$ (α_e)	0.039*	1.70	0.014*	2.82	0.013*	2.88
$\Lambda \to pe\nu$ (α_ν)	0.961*	5.42	0.975*	6.58	0.975*	6.60
$\Lambda \to pe\nu$ (α_p)	-0.586*	1.42	-0.579*	1.20	-0.579*	1.19
$(g_1/f_1)_{np}$	1.139*	373.00	1.255	0.00	1.250*	0.77
$(g_1/f_1)_{\Lambda p}$	0.654*	1.13	0.708	0.27	0.709*	0.31
$(g_1/f_1)_{\Sigma^- n}$	-0.317*	11.40	-0.387	1.85	-0.375*	3.12
$(f_1/g_1)_{\Sigma^-\Lambda}$	0.000*	0.34	0.000	0.24	0.000*	0.34

$F_{(a)} = 1.007$	$D_{(a)} = -1.329$	$\sin\theta_{(a)} = 0.221$	$C_{(a)} = 0.080 \pm 0.075$	$\chi^2_{(a)} = 6.83$
$F_{(b)} = 1.063$	$D_{(b)} = -1.499$	$\sin\theta_{(b)} = 0.228$	$C_{(b)} = -0.002 \pm 0.007$	$\chi^2_{(b)} = 10.92$
$F_{(c)} = 1.074$	$D_{(c)} = -1.481$	$\sin\theta_{(c)} = 0.227$	$C_{(c)} = 0.002 \pm 0.011$	$\chi^2_{(c)} = 8.29$

while the information on the rates and angular-correlation coefficients is in quite acceptably good agreement with the CT. Contrastingly, the fit with rates and g_1/f_1 ratios is satisfactory, which is misleading because it hides the deviations in the polarization data. It must be remarked that the predictions for the polarization data of $\Lambda \rightarrow pe\nu$ and $\Sigma^- \rightarrow ne\nu$ are the same as before. Evidently, the CT parameters and the different corrections we have considered are completely constrained by the experimental values of the transition rates and angular-correlation coefficients. The deviations between the polarization data and theory must be attributed to some other cause.

The main conclusion that we can draw from the above analysis is that the CT predictions are rendered stable by the transition rates and the $\alpha_{e\nu}$ coefficients and that there does not seem to be any major disagreement with this subset of data. Any addition to the HSD data will then provide a genuine test of the postulates of the CT. It is in the above sense that the CT obtains predictive power.

7.4. A Critical Analysis: Agreement and Disagreement.

So far we have only used the data of Table 3.1, which cover all the data of the latest Review of Particle Properties issue. There are more pieces of data that have been published after this last issue was put out. They are incorporated in Table 3.4

The reason why we have tacitly left these pieces of data out of the analysis of the last sections is that, as we shall see shortly, a very strong contradiction to the predictions of the CT at least in the form the theory was stated in Sec. 4.2 is obtained. When these pieces of data are included, the χ^2 is risen to a very high value and there exists the risk, as we have just mentioned at the end of Sec. 3, that the CT parameters may be poorly determined. Our analysis of what we called the working assumptions would be rendered unreliable. We now know that the CT parameters are fixed by rates and $\alpha_{e\nu}$ coefficients, making their values stable, and there is no question now that the corrections to replace the working assumptions are important. It is also clear that nothing miraculous can be expected from the latter. Incorporating the new pieces of information produces the results of Table 4. χ^2 grows up to 64.5 a change of 38 from its value of 26 in Table 2(c). It is more than doubled. In Table 4, the

TABLE 4. Comparison of CT with the most recent data, Table 3.4. Radiative corrections and the q^2-dependence of f_1 and g_1 are accounted for (see text).

	Prediction	$\Delta\chi^2$
$n \to pe\nu$ (rate)	1.087	0.18
$\Sigma^+ \to \Lambda e\nu$ (rate)	0.272	0.12
$\Sigma^- \to \Lambda e\nu$ (rate)	0.451	12.42
$\Lambda \to pe\nu$ (rate)	3.247	1.13
$\Sigma^- \to ne\nu$ (rate)	6.802	0.52
$\Xi^- \to \Lambda e\nu$ (rate)	2.925	4.57
$\Xi^- \to \Sigma^o e\nu$ (rate)	0.516	0.02
$\Lambda \to p\mu\nu$ (rate)	0.553	0.13
$\Sigma^- \to n\mu\nu$ (rate)	3.174	0.25
$\Xi^- \to \Lambda\mu\nu$ (rate)	0.833	0.37
$n \to pe\nu$ ($\alpha_{e\nu}$)	-0.073	0.06
$n \to pe\nu$ (α_e)	-0.080	2.24
$n \to pe\nu$ (α_ν)	0.989	0.12
$\Sigma^\pm \to \Lambda e\nu$ ($\alpha_{e\nu}$)	-0.408	0.15
$\Sigma^- \to \Lambda e\nu$ ($\alpha_{e\nu}$)	-0.412	0.03
$\Sigma^- \to ne\nu$ ($\alpha_{e\nu}$)	0.342	5.81
$\Sigma^- \to ne\nu$ (α_e)	-0.608	20.88
$\Lambda \to pe\nu$ ($\alpha_{e\nu}$)	-0.018	0.11
$\Lambda \to pe\nu$ (α_e)	0.010	3.03
$\Lambda \to pe\nu$ (α_ν)	0.977	6.72
$\Lambda \to pe\nu$ (α_p)	-0.578	1.16
$\Sigma^- \to \Lambda e\nu$ (A)	0.059	0.02
$\Sigma^- \to \Lambda e\nu$ (B)	0.900	0.51
$\Xi^- \to \Lambda e\nu$ ($\alpha_{e\nu}$)	0.651	1.47
$\Xi^- \to \Lambda e\nu$ (A)	0.462	2.48

$F = 1.104 \pm 0.014$ $D = -1.449 \pm 0.011$ $\sin\theta = 0.227 \pm 0.002$ $\chi^2 = 64.50$

model dependent part of radiative corrections is fixed at its estimate of Eq. (5.39) and the A-slopes at their dipole estimates. The A and B asymmetries of $\Sigma^- \rightarrow \Lambda e\nu$ and $\alpha_{e\nu}$ of $\Xi^- \rightarrow \Lambda e\nu$ are in good agreement with CT. It is the $\Sigma^- \rightarrow \Lambda e\nu$ rate and α_e in $\Sigma^- \rightarrow ne\nu$ that cause such a high rising of χ^2. Also, notice that although F, D, and θ have practically not changed from their values of Table 2(c), their changes do amount to several standard deviations.

We shall now concentrate on each of these two very significant deviations. Let us begin with the one from the $\Sigma^- \rightarrow \Lambda e\nu$ rate. It is clear that the high significance of the deviation of the CT prediction for this transition rate does not come from a major deviation between the central value and the theoretical prediction. It comes instead from the smallness of the new error bars. This then leads us to suspect that this discrepancy can be attributed to the presence of SU(3) symmetry-breaking effects. Indeed, the theoretical expression for the $\Sigma^- \rightarrow \Lambda e\nu$ rate is

$$R_{\Sigma-\Lambda} = \left[3.66\, f_1^2 + 10.95\, g_1^2 + 0.01\, f_2^2 + 0.03\, g_2^2 + 0.02\, f_1 f_2 - 1.00\, g_1 g_2\right] \cdot 10^5. \qquad (1)$$

where we have limited the coefficients to two decimal places and we have dropped the q^2-dependence of f_1 and g_1, because as we have seen earlier the change from this contributions is small enough, and it is how $g_1(0)$ —through D— is changed what is important. Since the CT predicts $f_1 = g_2 = 0$ and f_2 is kinematically suppressed, only the second term is relevant for the CT. If symmetry-breaking is allowed, one may expect that a g_2 as large as 20% of g_1^0 —the CT value— be present and a change of also 20% be induced in f_2. In contrast [2,3], f_1 would only change by some 4% of f_1^0 only. One can then estimate what change Δg_1 of g_1^0 is required to reproduce the world average for $\Sigma^- \rightarrow \Lambda e\nu$ rate. From Eq. (1), one can derive to first order in g_1,

$$\left(1 - 2\,\frac{\Delta g_1}{g_1^0}\right) \frac{10.95 \mp 0.20}{10.95} = \frac{0.377}{0.467}\,. \qquad (2)$$

In Eq. (2) we have dropped all terms of the order of a fraction of a percent and the double sign corresponds to the unknown relative phase between g_1 and g_2. It turns out that $\Delta g_1/g_1^0$ is 9% if the upper sign is chosen and 10% if the lower one is used. This change in g_1^0 could possibly be due to symmetry-breaking corrections.

One very interesting point to realize is that the best quark model estimates available[4] for Δg_1 in $\Sigma^- \to \Lambda e\nu$ predict it to be zero. It is very often the case that simplifying assumptions are introduced for no other reason than the lack of need of more refined approaches. It may well be the case of this recent quark model calculation. Donoghue and Holstein have discussed this situation in a very recent communication[5]. Their analysis, which is independent from ours[6], has led them to establish the above discrepancy with the CT from a different point of view. Clearly, the new value of the $\Sigma^- \to \Lambda e\nu$ rate may have significant theoretical relevance.

Let us now turn to the second discrepancy, that of α_e in $\Sigma^- \to ne\nu$. In this case, it is clear that this discrepancy is due not the size of the error bars but to the big separation between the experimental central value and the CT prediction. The best way to visualize the importance of this discrepancy is to revert to the graphical method we discussed in Sec. 3.2. Fixing the A-slopes at their pole-dominance value and incorporating the CVC prediction for the V-slopes we get the graph of Fig. 1. There we have drawn the old value of α_e in $\Sigma^- \to ne\nu$. We have also drawn the two narrow ranges that the current value of $|g_1/f_1|$ —once it is corrected by the q^2-dependence of form factors— allows for α_e. One should now put on top of this figure the current world average of α_e of 0.26 ± 0.19, which implies an upwards shift for α_e.

Considering α_e alone one would not believe how important a deviation from CT its present experimental value is, since it is only 4 standard deviations away. Seen from the point of view of $\alpha_{e\nu}$, it is another thing. In reality, all we need α_e to do for us is to tell us which sign g_1/f_1 has. The signs of α_e and g_1/f_1 should be the same. When α_e was at 0.04 this question was undecided. But now the odds are overwhelmingly that g_1/f_1 has positive sign, opposite to CT prediction. The upper "window" in Fig. 1 seems to be prefered.

For a final determination of g_1/f_1 it is nevertheless necessary not only that the central value of α_e be confirmed, but also that the error bars be much reduced. A non-zero g_2 could be present and α_e would then help in determining g_1/f_1 and g_2/f_1, simultaneously.

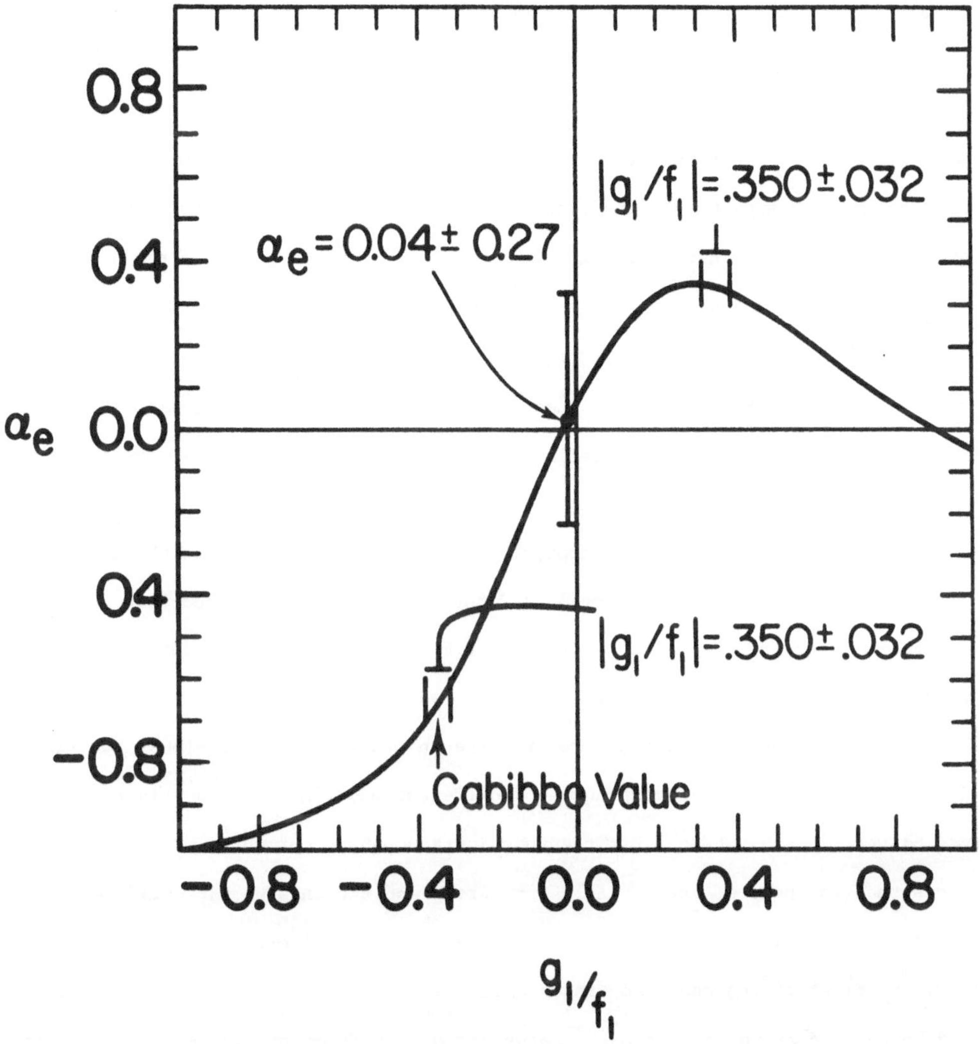

FIG. 1. The electron-spin asymmetry parameter α_e as a function of g_1/f_1 for $\Sigma^- \rightarrow ne^-\nu$. The data point is the weighted mean $\alpha_e = +0.04 \pm 0.27$ from Table 3.1. The "windows" allowed by the results of Ref. 3.10 are indicated by vertical lines.

Such a drastic change, as the selection of the opposite sign of g_1/f_1 is, amounts in $\Sigma^- \rightarrow ne\nu$ to a 200% breaking of SU(3) symmetry, because the magnitude of g_1/f_1 is big. If this happens to be finally true, a profound revision of CT would be required.

There are other deviations from CT. α_e and α_ν in $\Lambda \rightarrow pe\nu$ are worthwhile mentioning. Both show some tendency to deviate from the theoretical prediction. By looking at the graphs of α_e and α_ν against g_1/f_1 in $\Lambda \rightarrow pe\nu$, refered to in Sec. 3.2, a small change in α_e can be easily obtained by small changes in g_1/f_1, but a small change in α_ν can only be obtained with big changes in g_1/f_1. With current statistics one cannot yet say that even if α_ν is at the verge of three standard deviations from CT prediction it is a strong discrepancy. Nevertheless, refined measurements of α_ν in $\Lambda \rightarrow pe\nu$ are important since they could also bring a surprise.

At any rate, let us close this chapter with a word of caution. Both the above two strong discrepancies must be confirmed by future experiments. The fact that the most recent measurements of the corresponding experimental quantities are consistent with earlier ones is not yet sufficient, because the earlier measurements were all low statistics experiments. Another two remarks are in order. Our conclusions are valid as long as the experimental acceptances of different experiments do not introduce strong biases. This has been a tacit assumption throughout our work. Otherwise, world averages such as those of the Particle Data Group and the ones we use would not be very meaningful. Also, although we have given a very definite criteria to find deviations from CT, by first fixing and rendering stable the values of F, D, and θ, this point should be revised as the precision on rates and $\alpha_{e\nu}$ coefficients is increased. It is conceivable that at the end it is not the $\Sigma^- \rightarrow \Lambda e\nu$ rate but another one —e.g., the $\Xi^- \rightarrow \Lambda e\nu$ rate— that is indeed off the CT predictions.

Chapter 8. A Revision of Postulates: Symmetry Breaking and other Possibilities

8.1. First-Order Symmetry Breaking.

At this stage we have fully established the predictive power of CT, and we have studied in detail the relevance of the traditional approximations or working assumptions used to compare theory and experiment in HSD. The results of the last chapter clearly indicate that there is good agreement between CT and many pieces of data, mostly rates and $\alpha_{e\nu}$ coefficients, while there are deviations in some polarization asymmetries —α_e and α_ν in $\Lambda \to pe\nu$— and strong disagreements in two recent pieces of data —each around 4 standard deviations— namely, the $\Sigma^- \to \Lambda e\nu$ rate and the α_e in $\Sigma \to ne\nu$. Throughout this and the next two sections we incorporate radiative corrections, V- and A-slopes as in Sec. 7.4. Having under reasonable control the working assumptions, we must now turn[1] to questioning the postulates of CT if we want to understand the meaning of such deviations and disagreements.

Of the 6 postulates of CT, the weakest of them all is the last one, namely, the validity of the SU(3) symmetry limit. In order not to be too repetitive and at the risk of abusing abbreviations, we shall often refer to symmetry breaking by SB in this chapter. Certainly, one of the most attractive features of HSD is that they may provide clean experimental evidence on SU(3) SB, other than that coming from hyperon mass differences. As we remarked before, the CT was never intended to be exact and deviations from experiment are expected to appear. Therefore, one must first incorporate SB corrections to the CT before one may draw conclusions about its detailed success. Hence, it is most interesting to see how the predictions of the standard CT are changed once first-order SB is taken into account.

The straightforward way to proceed would be to obtain model calculations of SB, add them to the CT, and then compare it with experiment. Unfortunately, this kind of calculation is difficult to perform and only a few are available[7.4]. Because of the difficulties involved, the model calculations contain very particular assumptions and this kind of approach may not be as general as would be desirable. We shall take a different point of view.

We shall attempt to extract from experiment the SB corrections in a form which is as general as possible. This procedure may allow us to obtain the improved predictions of the CT. Assuming that SB comes from the eighth component of an octet in the strong-interaction Hamiltonian, one can obtain the most general first-order corrections to the axial-vector form factors g_1 and g_2. The SB contributions to g_1 would come from

$$A_1 \mathrm{Tr}(\{\lambda_i,\lambda_8\}\bar{B}B),\ B_1 \mathrm{Tr}(\bar{B}\{\lambda_i,\lambda_8\}B),$$

$$C_1\left[\mathrm{Tr}(\bar{B}\lambda_i B\lambda_8) - \mathrm{Tr}(\bar{B}\lambda_8 B\lambda_i)\right],$$

$$D_1\left[\mathrm{Tr}(\bar{B}\lambda_i)\mathrm{Tr}(B\lambda_8) + \mathrm{Tr}(\bar{B}\lambda_8)\mathrm{Tr}(B\lambda_i)\right],$$

$$E_1 \mathrm{Tr}(\bar{B}B)\mathrm{Tr}(\lambda_i\lambda_8), \tag{1}$$

and to g_2 would come from

$$A_2 \mathrm{Tr}([\lambda_i,\lambda_8]\bar{B}B),\ B_2 \mathrm{Tr}(\bar{B}[\lambda_i,\lambda_8]B),$$

$$C_2\left[\mathrm{Tr}(\bar{B}\lambda_i)\mathrm{Tr}(B\lambda_8) - \mathrm{Tr}(\bar{B}\lambda_8)\mathrm{Tr}(B\lambda_i)\right]. \tag{2}$$

This will lead to the following expressions for the axial-vector form factors g_1 and g_2 in terms of F, D, and the new reduced form factors $A_1,\ldots,D_1,A_2,\ldots,C_2$ for the SB contributions:

$$g_1(n\to p) = \frac{1}{\sqrt{6}}F - \left(\frac{3}{10}\right)^{1/2} D + \frac{2}{\sqrt{3}}(B_1 - C_1),$$

$$g_1(\Sigma^{\pm}\to\Lambda) = -\frac{1}{\sqrt{5}}D + \frac{\sqrt{2}}{3}(A_1 + B_1 + 3D_1),$$

$$g_1(\Lambda\to p) = -\frac{1}{2}F + \frac{1}{2\sqrt{5}}D + \frac{\sqrt{2}}{6}(-A_1 + 2B_1 + 3C_1 + 6D_1),$$

$$g_1(\Sigma^{-}\to n) = -\frac{1}{\sqrt{6}}F - \left(\frac{3}{10}\right)^{1/2} D - \frac{1}{\sqrt{3}}(A_1 + C_1),$$

$$g_1(\Xi^- \to \Lambda) = \frac{1}{2} F + \frac{1}{2\sqrt{5}} D + \frac{\sqrt{2}}{6} (2A_1 - B_1 - 3C_1 + 6D_1),$$

$$g_1(\Xi^- \to \Sigma^0) = \frac{1}{2\sqrt{3}} F - \frac{1}{2}\left(\frac{3}{5}\right)^{1/2} D + \frac{1}{\sqrt{6}} (-B_1 + C_1), \tag{3}$$

and

$$g_2(n \to p) = 0,$$

$$g_2(\Sigma^\pm \to \Lambda) = -\sqrt{2}\, C_2,$$

$$g_2(\Lambda \to p) = \sqrt{2}\left(-\frac{1}{2} A_2 + B_2 + C_2\right),$$

$$g_2(\Sigma^- \to n) = -\sqrt{3}\, A_2,$$

$$g_2(\Xi^- \to \Lambda) = \sqrt{2}\left(A_2 - \frac{1}{2} B_2 - C_2\right),$$

$$g_2(\Xi^- \to \Sigma^0) = -\left(\frac{3}{2}\right)^{1/2} B_2. \tag{4}$$

These pseudo-tensor form factors have the factor M_1, that was dividing them in Eq. (2.12), absorbed in them. This way we avoid a bias that is introduced by changing the normalization of g_2 from one decay to another. The parameters A_2, B_2, and C_2 have dimensions of GeV^{-1}. The change

$$\frac{1}{M_1} g_2 \to g_2 \tag{5}$$

should be understood in this and in the following two sections. Although there should be in all eight new contributions coming from Eqs. (1) and (2), in practice there are only seven. The reason is that the term $Tr(\bar{B}B)Tr(\lambda_i\lambda_8)$ is diagonal and cannot contribute to these matrix elements. The new quantities A_1, B_1, C_1, D_1, A_2, B_2, and C_2 in Eqs. (3) and (4) can be used to parametrize the SB in the experimental data, provided none of them becomes too large.

Before going into more detail, it is important to discuss further what is meant by first-order SB in the CT. Although one customarily says that the CT assumes that the symmetry limit is a good approximation, in reality SB is introduced to all orders into CT by keeping the physical masses of hyperons, since otherwise the available phase space would be zero and the decays would not take place at all. So what is really meant in the CT by the symmetry limit is that only the form factors are kept at their symmetry-limit values, while the difference ΔM between the hyperon masses is kept to all orders. Therefore, incorporating first-order SB into HSD in the spirit of the CT means that first-order corrections should be incorporated into $f_1(0)$ and $g_1(0)$ only, while f_2, g_2, and the slopes of the q^2-dependence of f_1 and g_1 should be kept at their symmetry-limit values. Strictly speaking, one should add all SB corrections in such a way that the order is well kept. For example, if the q^2-dependence of f_1 and g_1 is introduced, the second-order SB corrections to $f_1(0)$ and $g_1(0)$ should be included and first-order corrections to $f_2(0)$ and $g_2(0)$ should also be included and powers of $(\Delta M)^3$ and higher should be dropped; this would be a rigorous accounting of second-order SB in CT. We shall not pursue this approach in this paper. We shall instead stay close to the original spirit of the CT. Therefore, by first-order SB we mean that we shall keep f_2 at its CVC value, $g_2 = 0$ because of the absence of second-class currents, and f_1 at its symmetry-limit value because of the Behrends-Sirlin[7.2] and Ademollo-Gatto[7.3] theorems. Thus, only the changes in g_1 introduced in Eq. (3) should be considered.

8.2. Symmetry Breaking from Experiment.

Eqs. (1-4) provide the most general formulation of first-order SB for the observed processes in HSD within the CT and consistent with the Behrends-Sirlin and Ademello-Gatto theorem. It is this generality we want to exploit. In our opinion this is better than just using whatever estimates for SB in HSD are available, especially so, because the $\Sigma^- \to \Lambda e\nu$ rate is not supposed to be affected by SB in some models.

Fitting A_1, B_1, C_1, and D_1 along with F, D, and θ, we obtain the results (a) in Table 1. They are quite interesting. Comparing the values of these F, D, and θ with those of Table 7.2, we see that they changed very little, meaning that SB is indeed small. Also the values of A_1, B_1, C_1, and D_1 are small enough as to be acceptable as first-order SB. Looking at the new predictions we can observe that this pattern of SB leads to fine readjustments of some quantities. Indeed the rates for $\Sigma^+ \to \Lambda e\nu$ and $\Sigma^- \to \Lambda e\nu$ are reduced

TABLE 1. Comparison of the experimental data of Table 3.4 for HSD with CT with first-order symmetry breaking. (a) Incorporates the corrections in g_1 only, (b) incorporates the corrections in g_2 only, and (c) incorporates the corrections in g_1 and g_2. The corresponding values obtained for the parameters are given in Table 1(A), and the explicit values of the form factors for fit (a) are given in Table 1(B).

	(a)		(b)		(c)	
	Prediction	$\Delta\chi^2$	Prediction	$\Delta\chi^2$	Prediction	$\Delta\chi^2$
$n \to pe\nu$ (rate)	1.094	1.00	1.092	0.73	1.090	0.49
$\Sigma^+ \to \Lambda e\nu$ (rate)	0.234	0.06	0.238	0.04	0.236	0.05
$\Sigma^- \to \Lambda e\nu$ (rate)	0.389	0.00	0.392	0.07	0.389	0.00
$\Lambda \to pe\nu$ (rate)	3.180	0.01	3.269	2.10	3.170	0.01
$\Sigma^- \to ne\nu$ (rate)	6.995	0.00	6.615	2.46	6.946	0.00
$\Xi^- \to \Lambda e\nu$ (rate)	3.280	0.02	2.946	4.08	3.397	0.23
$\Xi^- \to \Sigma^0 e\nu$ (rate)	0.530	0.00	0.510	0.04	0.530	0.00
$\Lambda \to p\mu\nu$ (rate)	0.541	0.20	0.551	0.14	0.551	0.14
$\Sigma^- \to n\mu\nu$ (rate)	3.262	0.67	3.135	0.12	3.335	1.19
$\Xi^- \to \Lambda\mu\nu$ (rate)	0.932	0.32	0.830	0.37	0.935	0.31
$n \to pe\nu$ ($\alpha_{e\nu}$)	-0.074	0.01	-0.074	0.03	-0.075	0.01
$n \to pe\nu$ (α_e)	-0.081	0.67	-0.081	1.08	-0.082	0.37
$n \to pe\nu$ (α_ν)	0.989	0.13	0.989	0.13	0.989	0.13
$\Sigma^\pm \to \Lambda e\nu$ ($\alpha_{e\nu}$)	-0.408	0.15	-0.405	0.13	-0.401	0.12
$\Sigma^- \to \Lambda e\nu$ ($\alpha_{e\nu}$)	-0.412	0.04	-0.407	0.01	-0.402	0.00
$\Sigma^- \to ne\nu$ ($\alpha_{e\nu}$)	0.307	1.19	0.275	0.02	0.290	0.19
$\Sigma^- \to ne\nu$ (α_e)	-0.645	22.68	-0.547	18.04	-0.057	2.79
$\Lambda \to pe\nu$ ($\alpha_{e\nu}$)	-0.011	0.01	-0.002	0.67	-0.014	0.00
$\Lambda \to pe\nu$ (α_e)	0.014	2.85	0.021	2.47	0.018	3.48
$\Lambda \to pe\nu$ (α_ν)	0.975	6.58	0.982	7.23	0.948	4.45
$\Lambda \to pe\nu$ (α_p)	-0.579	1.20	-0.590	1.57	-0.552	0.46
$\Sigma^- \to \Lambda e\nu$ (A)	0.064	0.07	0.064	0.01	0.064	0.01
$\Sigma^- \to \Lambda e\nu$ (B)	0.901	0.54	0.900	0.52	0.899	0.49
$\Xi^- \to \Lambda e\nu$ ($\alpha_{e\nu}$)	0.514	0.03	0.665	1.82	0.704	3.03
$\Xi^- \to \Lambda e\nu$ (A)	0.634	0.02	0.546	0.55	0.661	0.17
χ^2		38.41		44.38		18.21

TABLE 1(A). Values obtained for the parameters of CT and of first-order symmetry breaking corrections from Eqs. (3) and (4). Notice that A_2, B_2, and G_2 have dimensions GeV^{-1}. See also Table 1(B).

$F = 1.235 \pm 0.152$	$F = 1.237 \pm 0.044$	$F = 1.463 \pm 0.191$
$D = -1.386 \pm 0.070$	$D = -1.355 \pm 0.033$	$D = -0.912 \pm 0.102$
$\sin\theta = 0.226 \pm 0.003$	$\sin\theta = 0.225 \pm 0.002$	$\sin\theta = 0.237 \pm 0.005$
$A_1 = -0.153 \pm 0.109$	A_1 -	$A_1 = 0.11 \pm 0.13$
$B_1 = -0.053 \pm 0.030$	B_1 -	$B_1 = 0.257 \pm 0.047$
$C_1 = -0.040 \pm 0.065$	C_1 -	$C_1 = 0.124 \pm 0.069$
$D_1 = 0.055 \pm 0.026$	D_1 -	$D_1 = 0.018 \pm 0.027$
A_2 -	$A_2 = 0.322 \pm 0.096$	$A_2 = 1.492 \pm 0.34$
B_2 -	$B_2 = -0.01 \pm 0.11$	$B_2 = 1.17 \pm 0.22$
C_2 -	$C_2 = -0.015 \pm 0.094$	$C_2 = -0.03 \pm 0.13$
$\chi^2 = 38.41$	$\chi^2 = 44.38$	$\chi^2 = 18.21$

TABLE 1(B). For completeness, we give in this table the explicit values of the form factors corresponding to the CT fit with first order SB corrections of Table 1(a), which is the best prediction of the CT (see later in the text). f_2 is normalized to m_p as explained in Sec. 4.3, and g_2, f_3, and g_3 are zero.

	f_1	g_1	f_2
np	1	1.249	1.853
$\Sigma^{\pm}\Lambda$	0	0.601	1.172
Λp	-1.225	-0.867	-1.098
$\Sigma^{-}n$	-1	0.366	1.017
$\Xi^{-}\Lambda$	1.225	0.354	-0.074
$\Xi^{-}\Sigma^{0}$	0.707	0.899	1.310

by ∿17% and the rates for $\Xi^- \rightarrow \Lambda e\nu$ and $\Xi^- \rightarrow \Lambda\mu\nu$ are increased by ∿10%. $\alpha_{e\nu}$ in $\Sigma^- \rightarrow ne\nu$ is decreased by ∿9% and α_e in $\Sigma^- \rightarrow ne\nu$ and A in $\Xi^- \rightarrow \Lambda e\nu$ are increased by ∿9% and ∿25%, respectively. All other changes are fairly small. The most dramatic change is in the rate for $\Sigma^- \rightarrow \Lambda e\nu$. It is changed by 3.6 standard deviations, and its experimental value is very well reproduced. But, the new value of α_e in $\Sigma^- \rightarrow ne\nu$ still remains totally off, as was to be expected.

Because of this we shall consider the incorporation of the g_2 form factors. Although it may not be easy to give an example, it is conceivable that for some reason SB may be stronger in the g_2 terms than elsewhere, and thus g_2 could be incorporated while keeping all other form factors as before. Besides it is interesting to see how big a g_2 is required by present data. In Table 1 we give two more cases (b) and (c). In (b) A_1, B_1, C_1, and D_1 are kept at zero and only A_2, B_2, and C_2 are allowed to vary and in (c) we take the combined effect of the corrections to g_1 and g_2. Except for minor details, fit (b) is very similar to fit (a). In contrast, fit (c) has a very much reduced χ^2, but this requires very large A_1, B_1, and D_1; too large to be considered a manifestation of first-order SB.

Also, the parameters A_2 and B_2 —if they are divided by a mass say around 1 GeV to make them dimensionless— are as big or even bigger than F and D. This makes g_2 as big as g_1^0 (the uncorrected g_1). Both the change in g_1 and the size of g_2 amount to a 200% SB correction. This is incompatible with the expansions of Eqs. (3) and (4).

Our main conclusion in this section is that small symmetry breaking through g_1 leads to very good agreement with present HSD data, except for the value of α_e in $\Sigma^- \rightarrow ne\nu$. Nevertheless, we can only claim that the incorporation of first-order SB into the CT is just consistent with present data. The approach of this section leads to noticeable modifications in the predictions of the original CT, which cannot yet be rigorously tested at present because of the laxity of some pieces of data. For example the substantial reduction predicted in the $\Sigma^+ \rightarrow \Lambda e\nu$ rate requires a much more precise measurement of such a rate. Clearly, with a substantial improvement of the data it will be possible to extract from the data, in a rather general, way important information on SB that would be of great use in guiding the theoretical work in this area.

8.3. Higher Representations.

By the title of this section we mean SU(3) representations for the axial-vector current other than the octet one. About the postulates of the CT, one may accept that beyond the postulate of the symmetry breaking the postulate of the choice of an octet representation for A_μ is not supported by much theory, as the CVC assumption for the V_μ case is. That V_μ is an octet suggests that A_μ be another one, but this may be more an assumption on simplicity than anything else. It is therefore[2] not idle to question if A_μ is really a pure SU(3) octet. We now shall put this question on less footing than the symmetry limit. That is, we shall assume that this limit is valid, even if A_μ may have mixed SU(3) transformation properties. One could also ask if both the symmetry limit and the octet assumptions are simultaneously violated. Unfortunately, we cannot test this possibility because HSD data are not precise nor abundant enough to do so. Nevertheless, that the symmetry limit may be saved by small admixtures of other representations for A_μ is, at least, a useful possibility.

Of course, the real motivation for challenging the octet A_μ current comes from the results of the last section. α_e in $\Sigma^- \to ne\nu$ was not affected by SB, even when non-zero g_2 was allowed in. The reason behind this was given already in Sec. 7.4, in Fig. 7.1. The new world average of this α_e combined with $\alpha_{e\nu}$ selects the opposite sign for g_1/f_1, which amounts to a 200% change in the symmetry limit prediction that CT does for it. It is therefore conceivable that it is the octet assumption for A_μ that needs revision and not the assumption on the validity of the symmetry limit.

Therefore, in addition to the octet, there may be 10, $\overline{10}$, and 27 in A_μ, i.e., A_μ is given by

$$A_\mu = A_\mu^{(8)} + A_\mu^{(10)} + A_\mu^{(\overline{10})} + A_\mu^{(27)}. \tag{6}$$

The Cabibbo universality assumes that the full weak current is obtained from the $\Delta S = 0$ current through a rotation around the 7th axis in the SU(3) space. We assume that this construction is also valid for higher representations. We thus have[1.1]

$$A_\mu = e^{-2i\theta_c F_7}\, \tilde{A}_\mu\, e^{2i\theta_c F_7}, \tag{7}$$

where

$$\tilde{A}_\mu = A_\mu^{(8)}(0,1,1) + A_\mu^{(10)}(0,1,1) + A_\mu^{(\overline{10})}(0,1,1) + A_\mu^{(27)}(0,1,1). \tag{8}$$

In Eq. (8), $A_\mu^{(n)}(Y,I,I_3)$ denotes the current that transforms according to the n representation of the SU(3) group and is the (Y,I,I_3) member of this representation. Eq. (8) indicates that in addition to the SU(3)-invariant form factors F and D, there will be three more reduced form factors: F_{10}, $F_{\overline{10}}$, and F_{27}. After performing the rotation (7) of the current (8) we obtain the following form of the weak axial-vector current:

$$\begin{aligned}
A_\mu = &-\left(\frac{5}{2}\right)^{1/2} \frac{1+\cos 2\theta_c}{2} \sin\theta_c A_\mu^{(27)}\left(-1,\frac{3}{2},\frac{3}{2}\right) + \frac{\sqrt{5}}{4}(1-\cos 2\theta_c)\cos\theta_c A_\mu^{(27)}(0,2,1) \\
&+ \frac{1}{4}(5\cos 2\theta_c - 1)\cos\theta_c A_\mu^{(27)}(0,1,1) - \left(\frac{5}{6}\right)^{1/2} \sin\theta_c \cos 2\theta_c A_\mu^{(27)}\left(1,\frac{3}{2},\frac{1}{2}\right) \\
&+ \frac{1}{2\sqrt{6}} \sin\theta_c (3+5\cos 2\theta_c) A_\mu^{(27)}\left(1,\frac{1}{2},\frac{1}{2}\right) - \left(\frac{5}{2}\right)^{1/2} \frac{1-\cos 2\theta_c}{2} \cos\theta_c A_\mu^{(27)}(2,1,0) \\
&- \sqrt{3}\, \frac{1+\cos 2\theta_c}{2} \sin\theta_c A_\mu^{(\overline{10})}\left(-1,\frac{3}{2},\frac{3}{2}\right) + \frac{3\cos 2\theta_c - 1}{2} \cos\theta_c A^{(\overline{10})}(0,1,1) \\
&+ \frac{3\cos 2\theta_c + 1}{2} \sin\theta_c A_\mu^{(\overline{10})}\left(1,\frac{1}{2},\frac{1}{2}\right) + \sqrt{3}\, \frac{1-\cos 2\theta_c}{2} \cos\theta_c A_\mu^{(\overline{10})}(2,0,0) \\
&+ \cos\theta_c A_\mu^{(10)}(0,1,1) - \sin\theta_c A_\mu^{(10)}\left(1,\frac{3}{2},\frac{1}{2}\right) + \cos\theta_c A_\mu^{(8)}(0,1,1) + \sin\theta_c A^{(8)}\left(1,\frac{1}{2},\frac{1}{2}\right).
\end{aligned} \tag{9}$$

The matrix elements of the current (9) can be calculated with the help of the Wigner-Eckart theorem. The corresponding Clebsch-Gordan coefficients[4.2] for the observed decays are compiled in Table 2.

TABLE 2. SU(3) Clebsch-Gordan coefficients for higher representations in HSD. The normalizations are as in Ref. (4.2).

	10	$\overline{10}$	27,(I)	27,(I)
$n \to p$	$-\left(\frac{2}{15}\right)^{1/2}$	$\left(\frac{2}{15}\right)^{1/2}$	$\frac{2}{3}\left(\frac{2}{15}\right)^{1/2}$, (1)	0, (2)
$\Sigma^{\pm} \to \Lambda$	$\frac{1}{\sqrt{5}}$	$\frac{1}{\sqrt{5}}$	$-\frac{2}{3\sqrt{5}}$, (1)	0, (2)
$\Lambda \to p$	0	$-\frac{1}{\sqrt{5}}$	$-\left(\frac{2}{15}\right)^{1/2}$, $\left(\frac{1}{2}\right)$	0, $\frac{3}{2}$
$\Sigma^{-} \to n$	$\left(\frac{2}{15}\right)^{1/2}$	$\left(\frac{2}{15}\right)^{1/2}$	$-\frac{2}{9\sqrt{5}}$, $\left(\frac{1}{2}\right)$	$-\frac{2}{9}$, $\left(\frac{3}{2}\right)$
$\Xi^{-} \to \Lambda$	0	$\frac{1}{\sqrt{5}}$	$-\left(\frac{2}{15}\right)^{1/2}$, $\left(\frac{1}{2}\right)$	0, $\left(\frac{3}{2}\right)$
$\Xi^{-} \to \Sigma^{0}$	$\frac{2}{\sqrt{15}}$	$-\frac{1}{\sqrt{15}}$	$-\frac{1}{9}\left(\frac{2}{5}\right)^{1/2}$, $\left(\frac{1}{2}\right)$	$\frac{2}{9}\sqrt{2}$, $\left(\frac{3}{2}\right)$

In the case of higher representations we have performed several fits. We have first tried separately each of the higher representations, then combinations of each with $A_{\mu}^{(8)}$, and finally all of them together. In neither one of all these options was there any improvement with respect to α_e of $\Sigma^- \to ne\nu$ found. Therefore we shall only display the results for the last case, they are given in Table 3. This table shows that the discrepancy for the rate of $\Sigma^- \to \Lambda e\nu$ disappears, while the deviation of α_e of $\Sigma^- \to ne\nu$ is not improved at all. The values of form factors F_{10}, $F_{\overline{10}}$, and F_{27} are small in comparison with the values of F and D, and these latter two do not differ significantly from the values in Table 7.2. This scheme cannot be discriminated from the first-order symmetry-breaking schemes of Table 2. We thus see that the presence of higher representations in A_{μ} cannot explain the new world average for α_e in $\Sigma^- \to ne\nu$ either.

8.4. Presence of other Flavors.

With the advent of gauge theories and the discovery of new quantum numbers like charm and bottom, it has become clear that the underlying theory supporting the effective V-A theory is more subtle and elaborate that what could have been expected

TABLE 3. Comparison of the experimental data of Table 3.4 for HSD with CT with higher representations of SU(3) allowed for the axial-vector current. The parameters F, D, F_{10}, $F_{\overline{10}}$, and F_{27} are the SU(3) reduced form factors of the axial vector current.

$n \to pe\nu$ (rate)	1.094	0.99
$\Sigma^+ \to \Lambda e\nu$ (rate)	0.234	0.06
$\Sigma^- \to \Lambda e\nu$ (rate)	0.389	0.00
$\Lambda \to pe\nu$ (rate)	3.180	0.01
$\Sigma^- \to ne\nu$ (rate)	6.992	0.02
$\Xi^- \to \Lambda e\nu$ (rate)	3.279	0.03
$\Xi^- \to \Sigma^o e\nu$ (rate)	0.547	0.03
$\Lambda \to p\mu\nu$ (rate)	0.541	0.20
$\Sigma^- \to n\mu\nu$ (rate)	3.260	0.66
$\Xi^- \to \Lambda\mu\nu$ (rate)	0.931	0.32
$n \to pe\nu$ ($\alpha_{e\nu}$)	-0.074	0.00
$n \to pe\nu$ (α_e)	-0.081	0.69
$n \to pe\nu$ (α_ν)	0.989	0.13
$\Sigma^+ \to \Lambda e\nu$ ($\alpha_{e\nu}$)	-0.408	0.15
$\Sigma^- \to \Lambda e\nu$ ($\alpha_{e\nu}$)	-0.412	0.04
$\Sigma^- \to ne\nu$ ($\alpha_{e\nu}$)	0.307	1.14
$\Sigma^- \to ne\nu$ (α_e)	-0.646	22.71
$\Lambda \to pe\nu$ ($\alpha_{e\nu}$)	-0.012	0.01
$\Lambda \to pe\nu$ (α_e)	0.013	2.87
$\Lambda \to pe\nu$ (α_ν)	0.975	6.60
$\Lambda \to pe\nu$ (α_p)	-0.579	1.19
$\Sigma^- \to \Lambda e\nu$ (A)	0.064	0.07
$\Sigma^- \to \Lambda e\nu$ (B)	0.902	0.54
$\Xi^- \to \Lambda e\nu$ ($\alpha_{e\nu}$)	0.513	0.03
$\Xi^- \to \Lambda e\nu$ (A)	0.634	0.02

$F = 1.125 \pm 0.019$ $\quad D = -1.363 \pm 0.024$ $\quad \sin\theta = 0.225 \pm 0.003$ $\quad \chi^2 = 38.44$

$F_{\overline{10}} = 0.058 \pm 0.025$ $\quad F_{10} = -0.118 \pm 0.023$ $\quad F_{27} = -0.086 \pm 0.031$

in the early days of the CT.

The sole existence of the Cabibbo angle has been questioned. That is, the form given to universality in the CT may have to be revised in the light of the existence of new quantum numbers. In a model with more than four quarks it is conceivable that the universality of weak interactions is modified, as illustrated by the model of Kobayashi and Maskawa[3]. This, of course, would be a change that should not be attributed to symmetry-breaking effects. Following a notation similar to Ref. (4), we put $V_{11} = \cos\theta$ and we replace $\sin\theta$ by another parameter $V_{12} = \sin\theta\cos\theta_1$, where θ_1 is another Cabibbo-type angle independent of θ. A modification of the CT universality could be detected by handling V_{11} and V_{12} as two independent parameters.

We shall use the data of Table 3.1, because the large deviations of the $\Sigma^- \to \Lambda e\nu$ rate and α_e of $\Sigma^- \to ne\nu$ force χ^2 so high, when no SB is accounted for, that we would risk that the fitting procedure be fouled by them. At any rate, the deviation from universality is hard to see in HSD and it should be established better with those pieces of data that better agree with CT. We shall then perform an analysis in the spirit of Sec. 7.4. It is clear that the role of C in changing G_μ as in Eq. (5.41) requires

$$G_V' = G_V(1+C), \tag{10}$$

which has an effect on universality which should not be confused with relaxing universality at all.

There is no need to reproduce the predictions for all the observables, because there is no apparent change. Thus, we have listed in Table 4 only the values obtained for V_{11}, V_{12}, F, and D for fits where the data are subdivided into transition rates and angular-correlation coefficients only and when all the data are used. We have performed fits under different assumptions, as explained in the caption to Table 4. To see an effect attributable to the presence of $\cos\theta_1$ it is required that $V_{11}^2 + V_{12}^2 < 1$. Going through Table 4, it can be seen that when C = 0 there is very small such effect, even after the incorporation of the V- and A-slopes (variable λ_F still gives negative A-slope for $\Xi^- \to \Lambda e\nu$). Only when C = 1% and when pole dominance A slopes are used an effect in the correct direction appears, fits (G) and (H). But when dipole A slopes are used the situation is less clear, fits (I) and (J). Apparently, if only the subset of data that better agrees with the CT is used one cannot see any significant deviation from the Cabibbo universality One might say that in order to see, in such a sector, the effect of $\cos\theta_1$, further symmetry-breaking must be introduced. The deviation from Cabibbo universality seen when all the data are used cannot be taken too seriously, because the polarization data in $\Lambda \to pe\nu$ and $\Sigma^- \to ne\nu$, which have been showing deviations from the CT all along, were not better fit and remain just as before.

Nevertheless, there is room enough for the effect of $\cos\theta_1$ to be present. This is only due to the current precision of the data. The error bars on V_{11} and V_{12} are 1.3% and 1.4%, respectively, if only rates and correlation coefficients are used, and 1% and 1.4% if all the data are used. Therefore we cannot conclude that $\cos\theta_1$ is excluded.[4]

To close this section, let us remark that in HSD only the combinations $G'_V\cos\theta$ and $G'_V\sin\theta\cos\theta_1$ can be determined experimentally. Hence, the appearance of $\cos\theta_1$ is intimately related to the value of C. So, within HSD the existence of $\cos\theta_1$ can only be solved theoretically by requiring a very reliable estimation of C.

TABLE 4. Fitted values of V_{11} and V_{12}. (A) and (B) correspond to using rates and angular-correlation coefficients only and all of the data, respectively, assuming no V- or A-slopes present and C = 0. In all the following fits CVC V-slopes are incorporated. In (C) and (D) the data are partitioned as in (A) and (B) respectively, assuming C = 0 and pole-dominance A-slopes. (E) and (F) are as (C) and (D), but with dipole A-slopes. The last four fits assume C = 1%. Otherwise, (G) and (H) are as (C) and (D) with pole-dominance A-slopes; and (I) and (J) are as (E) and (F) with dipole A-slopes, respectively. The quantities in parentheses are the percentage change from 1 in $V_{11}^2 + V_{12}^2$.

	F	D	V_{11}	V_{12}	$V_{11}^2 + V_{12}^2$	χ^2
(A)	1.062	-1.520	0.965	0.232	0.985 (-1.51%)	14.69
(B)	1.058	-1.504	0.972	0.233	0.998 (-0.17%)	32.82
(C)	1.079	-1.490	0.972	0.228	0.997 (-0.35%)	9.56
(D)	1.080	-1.458	0.973	0.228	0.999 (-0.09%)	26.67
(E)	1.074	-1.481	0.976	0.228	1.004 (0.42%)	8.39
(F)	1.079	-1.484	0.974	0.227	1.000 (-0.02%)	26.00
(G)	1.079	-1.490	0.962	0.226	0.977 (-2.31%)	9.56
(H)	1.080	-1.485	0.964	0.226	0.980 (-2.05%)	26.67
(I)	1.074	-1.481	0.966	0.225	0.984 (-1.57%)	8.39
(J)	1.079	-1.484	0.964	0.225	0.980 (-1.99%)	26.00

8.5. Second-Class Axial-Vector Currents.

In the hierarchy of the postulates of the CT, as we stated them in Sec. 4.2, we stressed the importance of the assumption that only first-class currents exist. The electromagnetic current is first-class and any model that attemps to unify weak and electromagnetic interactions will run into many difficulties if second-class weak currents exist. The present world average value of α_e in $\Sigma^- \to ne\nu$ is so far off from the CT prediction that the question arises if it could signal the existence of second-class A_μ currents.

We shall now relax[5] the first postulate of CT for A_μ; namely, that A_μ has a definite G-parity behavior. A_μ then would no longer be either a first or a second-class operator. Otherwise the CT stands as before. The hadronic matrix element in Eq. (2.12) requires then that g_2 be present. Since A_μ is still an octet and the symmetry limit is valid, g_2 is described by

$$g_2^{AB} = C_F^{AB} F_4 + C_D^{AB} D_4,$$

where F_4 and D_4 are new reduced form factors. The number of parameters is now five; they include the three CT parameters, F_4 and D_4.

We shall first use the data of Table 3.1, for the same reason as in last section, i.e., that χ^2 do not rise too high, and we shall afterwards replace the new world average of α_e in $\Sigma^- \to ne\nu$ instead of the old one.

Since our attention is focused on this α_e, we shall not reproduce complete tables of predictions. It is sufficient to quote the value obtained for it. Radiative corrections and V- and A-slopes are incorporated as in Sec. 1. The two fits essentially agree in everything. In both of them the new value of α_e in $\Sigma^- \to ne\nu$ is reduced to -0.60. When the old world average is used, the corresponding $\Delta\chi^2$ is 5.6. When the new world average is used $\Delta\chi^2$ simply becomes 20.5, still more than 4 standard deviations. The change from the prediction of the standard CT of -0.62 is too small.

We can, therefore, conclude that the current experimental value of α_e in $\Sigma^- \to ne\nu$ cannot be interpreted as the signal for second-class A_μ currents. If indeed these currents are to exist then their contribution should be very small. In this respect, we reach for HSD the same conclusion that was reached already in nuclear physics searches for second-class currents.

8.6. Discussion. The Best Prediction of Cabibbo Theory.

From the revision of the postulates of the CT that we have systematically performed in this chapter, we can conclude that only two extensions of the CT are meaningful —given the current experimental situation of HSD. The first one —a natural one, incorporating first-order SB— improves the agreement with experiment significantly and the rate of $\Sigma^- \to \Lambda e\nu$ can be explained very well. The second one —the admixtures of higher representations to A_μ — works just as well as the first one. But, in both extensions α_e of $\Sigma^- \to ne\nu$ deviates very much from its current experimental value. It then remains as a strong discrepancy.

We can summarize our conclusion of this chapter in terms of two mutually exclusive statements, depending on the future experimental value of α_e in $\Sigma^- \to ne\nu$:

(1) If the value of $\alpha_e = 0.26 \pm 0.19$ in $\Sigma^- \to ne\nu$ persists, then the CT will need some essential modification.

(2) If the value of α_e in $\Sigma^- \to ne\nu$ will move in the direction closer to the CT-favored value, then the only modification of CT that is required is first-order symmetry breaking. And, then, future more detailed experimental results on HSD would decide on the particular SB scheme.

The alternative of higher representations for A_μ provides no essential advantage of its own if it is only going, so to speak, mimic the role of first-order SB. From this point of view, we can say that relaxing the symmetry limit postulate of CT —with implications that can be observed through experiment— is a most welcome option. Hence, we can say that the best prediction of CT for HSD is the one of Table 1(a). The discrepancy with α_e in $\Sigma^- \to ne\nu$ would then be expected to be corrected by future experiments.

This is as far as we can presently go with CT to describe HSD. We shall now turn to study the consequences of α_e in $\Sigma^- \to ne\nu$ from another point of view, which addresses itself to the question how internal symmetries should be formualted for semileptonic decays, but which is otherwise very much in the spirit of CT. This will show how important an accurate measurement of α_e in $\Sigma^- \to ne\nu$ is.

Chapter 9. Spectrum Generating SU(3)

9.1. Exact Symmetry Formulation.

From the analysis of last chapter we are faced with the unique conclusion that, if the current world average of α_e in $\Sigma^- \to ne\nu$ is confirmed by future experiments, then strong symmetry-breaking must be present in HSD. Such a strong symmetry-breaking might be very difficult to compute perturbatively and, therefore, a different approach that redefines the concept of SU(3) symmetry might be required altogether. The only approach of this kind known to us that is in agreement with the present value of α_e in $\Sigma^- \to ne\nu$ is the approach that treats SU(3) as a spectrum generating group (SG)[1].

The central idea behind SG is to treat symmetry breaking, as it shows through hadron mass differences, in a quantum-mechanically consistent way. From the basic postulates of quantum mechanics, we know that from the outset we must choose a complete system of commuting operators. That is, the quantum-mechanical states that describe the particles of the problem we have in mind must be simultaneous eigenstates of the operators that correspond to the observables of such problem. In HSD this means that, when SU(3) is used, we should start by requiring that

$$[P_\mu, E_\alpha] = 0, \tag{1}$$

where P_μ are the four-momentum operators ($\mu = 0,1,2,3$) and E_α are the SU(3) generators ($\alpha = \lambda,\ldots,8$), in the Cartan notation. The space of states is therefore the direct-product space of SU(3) eigenstates and Poincaré eigenstates. However, assumption (1) is contradicted by experiment. The non-zero hadronic mass-differences imply that one cannot perform SU(3) rotations without affecting the Poincaré states. Thus, one has

$$[P_\mu, E_\alpha] \neq 0, \tag{2}$$

and one cannot use the Wigner-Eckart theorem anymore to write down matrix elements in terms of Clebsch-Gordan coefficients and reduced form factors while keeping Poincaré eigenstates. The way this problem is dealt with in the CT is to assume that (1) can be used as an approximation good to lowest order, and at a later stage one may introduce symmetry-breaking corrections perturbatively. This has been the logics he have followed all along. However, in view of the current value of α_e in $\Sigma^- \to ne\nu$ this approach is really subject to strong suspicion. Our basic quantum mechanics postulates then tell us that inequality (2) must not be ignored anymore if we do not have a reliable perturbation expansion to cope with it.

We are then faced with having to find a set of commuting operators from the outset. In other words, we should formulate SU(3) as an exact symmetry group despite the hadron mass differences. One choice is to propose[2] ("hats" in this chapter refer to velocity eigenstates only)

$$[\hat{P}_\mu, E_\alpha] = 0, \tag{3}$$

instead of Eq. (1), where $\hat{P}_\mu$ are the four-velocity operators. This choice is by no means unique, but it is certainly a simple and straight forward one. The physical requirement that the mass operator must not commute with the SU(3) raising and lowering operators is, however, rather restrictive and rules out many alternatives, such as $[P_\mu/M^2, E_\alpha] = 0$, to mention one example. The merit behind Eq. (3) is that it does not obviously contradict experiment and that it is not against any fundamental theoretical notion. The detailed experimental validity of Eq. (3) can be put to trial in HSD.

In order to avoid confusion, we shall be more careful with notation in this chapter[3]. The SU(3) indeces for A and B in the process $A \to Be\nu$ will be α and α', respectively. Their masses will be accordingly denoted by m_α and $m_{\alpha'}$. The index β will refer to the SU(3) indeces of the current operators and the index $\gamma = 1,2$ will mean antisymmetric or symmetric reduced form factors (indeces F and D of Sec. 4.3, respectively). C will still mean Clebsch-Gordan coefficients. The states of A and B will be denoted with kets and bras with the corresponding four-velocity and spin σ showing in them, when the matrix elements of the currents A_μ and V_μ are calculated.

To analize the consequences of assumption (3) it is most convenient, although not necessary, to employ the Poincaré group $P \equiv P(\hat{P}_\mu, L_{\mu\nu})$ generated by the four-velocity $\hat{P}_\mu$, rather than the physical Poincaré group $P(P_\mu, L_{\mu\nu})$ because this Poincaré group commutes with SU(3) and $\hat{P}_\mu \hat{P}^\mu = 1$. The mass operator M is then defined by its SU(3) transformation properties and the mass m_α (the matrix elements of this operator) is a function of isospin and hypercharge, etc., $m_\alpha = m(I,I_3,Y)$. Under assumption (3) we may write the basis vectors of the representation space $\mathcal{P}(P_\mu, L_{\mu\nu}) \times SU(3)$ as direct products of the form

$$|\hat{p}\sigma\alpha\rangle \equiv |\hat{p}\sigma\rangle \times |\alpha\rangle \ , \tag{4}$$

where $\{|\hat{p}\sigma\rangle\}$ span the respresentation space of $P(\hat{P}_\mu, L_{\mu\nu})$ ($\hat{p}$-space) and $\{|\alpha\rangle\}$ span the representation space of SU(3). This last we choose as an octet space. These states are to be properly normalized. The rigorous calculation details are explained in Ref. (3). The relevant matrix element of current operators with energy-momentum conservation is

$$\langle \hat{p}'\sigma'\alpha' | V^\beta_\mu + A^\beta_\mu | \hat{p}\sigma\alpha\rangle =$$

$$= (m_\alpha m_{\alpha'})^{3/2}\, \bar{u}(\hat{p}'\sigma') \{ F_1^{\alpha'\beta\alpha}(\hat{q}^2)\gamma_\mu + F_2^{\alpha'\beta\alpha}(\hat{q}^2)\sigma_{\mu\nu}\hat{q}_\nu + F_3^{\alpha'\beta\alpha}(\hat{q}^2)\hat{q}_\mu +$$

$$+ [G_1^{\alpha'\beta\alpha}(\hat{q}^2)\gamma_\mu + G_2^{\alpha'\beta\alpha}(\hat{q}^2)\sigma_{\mu\nu}\hat{q}_\nu + G_3^{\alpha'\beta\alpha}(\hat{q}^2)\hat{q}_\mu]\gamma_5\}\, u(\hat{p}\sigma) \tag{5}$$

This is exactly the same as for the momentum basis vectors except that p is replaced by $\hat{p}$ and $\hat{q} = \hat{p}-\hat{p}'$ is the four-velocity transfer between the hyperons. The upper-case letters for form factors serve as a reminder (they should not be confused at all with those of Sec. 2.3). The Wigner-Eckart theorem can be applied to these form factors without any reservations because they are SU(3)-covariant functions of the SU(3) invariant variable $\hat{q}$. For octet V_μ and A_μ, one has

$$F_i^{\alpha'\beta\alpha}(\hat{q}^2) = \sum_{\gamma=1,2} C(\gamma;\alpha'\beta\alpha) F_i^{\gamma}(\hat{q}^2), \tag{6}$$

and

$$G_i^{\alpha'\beta\alpha}(\hat{q}^2) = \sum_{\gamma=1,2} C(\gamma;\alpha'\beta\alpha) G_i^{\gamma}(\hat{q}^2) \tag{7}$$

The transition matrix element for HSD in the SG approach is given by

$$M = \frac{G_V}{\sqrt{2}} \bar{u}_e(\ell)\gamma_\mu(1+\gamma_5)v_\nu(p_\nu)$$

$$\times < \hat{p}'\sigma'\alpha' \mid J_\mu^\beta \mid \hat{p}\sigma\alpha > \tag{8}$$

where J_μ^β is $V_\mu^\beta + A_\mu^\beta$ or certain combination of them. For example, incorporating Cabibbo universality one should identify

$$J_\mu^{1+i2} = \cos\theta\,(V_\mu^{1+i2} + A_\mu^{1+i2}), \tag{9}$$

and

$$J_\mu^{4+i5} = \sin\theta\,(V_\mu^{4+i5} + A_\mu^{4+i5}). \tag{10}$$

It should be evident that the SG approach is perfectly consistent with the spirit of the original CT, except that the internal symmetry postulates are formulated in a rigorous way and not in an approximate one.

What remains to establish the applicability of the SG scheme is to connect the matrix element of Eq. (8) with M_0 of Eq. (2.11). Since the Poincaré group $P(\hat{P}_\mu, L_{\mu\nu})$ is just as well defined as $P(P_\mu, L_{\mu\nu})$ and the connection between the two is already determined by the connection between $\hat{P}_\mu$ and P_μ, the ordinary matrix element of HSD of Eq. (2.11) is determined, since the form factors of Eq. (2.12) become functions of the upper-case ones of Eq. (5). For octet V_μ and A_μ, we get[3]

$$f_1^{\alpha'\beta\alpha} = \sum_{\gamma=1,2} C(\gamma;\alpha\beta\alpha') \left[F_1^{(\gamma)} + \left(2 - \frac{(m_\alpha + m_{\alpha'})^2}{2m_\alpha m_{\alpha'}} \right) F_2^{(\gamma)} - \frac{m_\alpha^2 - m_{\alpha'}^2}{2m_\alpha m_{\alpha'}} F_3^{(\gamma)} \right],$$

$$f_2^{\alpha'\beta\alpha} = \frac{m_\alpha}{2(m_\alpha m_{\alpha'})} \sum_{\gamma=1,2} C(\gamma;\alpha\beta\alpha') \left[(m_\alpha + m_{\alpha'}) F_2^{(\gamma)} + (m_\alpha - m_{\alpha'}) F_3^{(\gamma)} \right],$$

$$f_3^{\alpha'\beta\alpha} = \frac{m_\alpha}{2(m_\alpha m_{\alpha'})} \sum_{\gamma=1,2} C(\gamma;\alpha\beta\alpha') \left[(m_\alpha - m_{\alpha'}) F_2^{(\gamma)} + (m_\alpha + m_{\alpha'}) F_3^{(\gamma)} \right],$$

$$g_1^{\alpha'\beta\alpha} = \sum_{\gamma=1,2} C(\gamma;\alpha\beta\alpha') \left[G_1^{(\gamma)} + \frac{m_\alpha^2 - m_{\alpha'}^2}{2m_\alpha m_{\alpha'}} G_2^{(\gamma)} + \frac{(m_\alpha - m_{\alpha'})^2}{2m_\alpha m_{\alpha'}} G_3^{(\gamma)} \right],$$

$$g_2^{\alpha'\beta\alpha} = \frac{m_\alpha}{2(m_\alpha m_{\alpha'})} \sum_{\gamma=1,2} C(\gamma;\alpha\beta\alpha') \left[(m_\alpha + m_{\alpha'}) G_2^{(\gamma)} + (m_\alpha - m_{\alpha'}) G_3^{(\gamma)} \right],$$

$$g_3^{\alpha'\beta\alpha} = \frac{m_\alpha}{2(m_\alpha m_{\alpha'})} \sum_{\gamma=1,2} C(\gamma;\alpha\beta\alpha') \left[(m_\alpha - m_{\alpha'}) G_2^{(\gamma)} + (m_\alpha + m_{\alpha'}) G_3^{(\gamma)} \right], \qquad (11)$$

where we have indicated the SU(3) indeces explcitly on the ordinary form factors.

In Eqs. (11) one must remark that a non-zero g_2 is generated even when $G_2^\gamma = 0$. If hyperon masses were equal then g_2 would be zero unless $G_2 \neq 0$. Another feature of these equations is that they naturally obey the Behrends-Sirlin[7.2] and Ademollo-Gatto[7.3] theorem in the absence of second-class operators. There are no corrections to first order in symmetry breaking to the vector form factors f_1 and f_2. The mass differences appear quadratically in them. As a matter of fact, the Behrends-Sirlin form of this theorem is more restrictive than the Ademollo-Gatto form. It states that in form factors of a conserved vector current the mass difference of the hadrons involved cannot appear to any odd power, while the sum of the masses can appear to any power. This is exactly what happens in f_1 and f_2 when second-class operators are absent.

9.2. Spectrum Generating Models.

Having formulated SU(3) as an exact internal symmetry, we must now proceed to introduce detailed assumptions. The first one is already implicit: the V-A theory is valid. The next one is, in close parallelism to the postulates of the CT of Sec. 4.2, the absence of second-class current operators. This implies that

$$F_3^{\gamma} = 0,$$
$$G_2^{\gamma} = 0, \quad \gamma = 1,2. \tag{12}$$

Next, there is that V_μ is an octet operator. This, of course suggests that the CVC assumption is valid for it. Accepting it, we can fix all the vector form factors. In the convention[4.2] for the Clebsch-Gordan coefficients we are using, we get

$$F_1^1 = \sqrt{6} \,,$$
$$F_1^2 = 0 \,,$$
$$F_2^1 = \sqrt{6}\left(\frac{\mu_p}{2} + \frac{\mu_n}{4}\right) ,$$
$$F_2^2 = \mu_n \sqrt{30}/4 \,, \tag{13}$$

where, just as in Eqs. (4.6)-(4.9), μ_p and μ_n are the anomalous magnetic moments of the proton and the neutron, respectively. For the A_μ current we shall try two possibilities. First that it is a pure octet and next -- in analogy to Ch. 8.3 -- that it may have in addition admixtures of higher representations, as we did with the CT in Sec. 8.3. We shall not consider any formulation of PCAC. In this respect SC models have two more parameters than the original CT, but the same if PCAC is not used as an extra assumption in CT, as we discussed in Sec. 6.2. The difference now is that G_3 shows also in the e-modes and not only in the μ-modes as was the case for g_3.

The last remaining assumption is about universality. We certainly can follow the CT, as was indicated in Eqs. (9) and (10), and this we will do. But, other options are open to us in SG. They have their own interest and should be, if not fully discussed, at least illustrated. As is well known the Cabibbo angle is a phenomenological parameter that has remained unrelated to fundamental constants and other physical quantities.

The efforts to relate it to symmetry-breaking and in particular, the more ambitious goal, to hadron mass differences have not led too far yet. As we stressed in the last section, the V_μ and A_μ operators act in $\hat{p}$-space; in going to the ordinary p-space we have some freedom. Instead of identifying V_μ and A_μ with the operators in p-space, we can identify them with other operators that involve the mass operator. For example, we can try[4]

$$J_\mu = \hat{J}_\mu + A_1[M,[M,\hat{J}_\mu]] + A_2[M,[M,[M,[M,\hat{J}_\mu]]]] + \ldots \tag{14}$$

$\hat{J}_\mu$ is the current operator in $\hat{p}$-space and J_μ is already in p-space. The equality sign here should be understood in the sense that the states with $\hat{p}$ go to states with p as in the direct product $|p\rangle = |\hat{p}\rangle \times |m\rangle$. In Eq. (14) only an even number of commutators can appear if time reversal is valid and the hermitian conjugate current J_μ^+ is the one to appear in $\Sigma^+ \to \Lambda e\nu$. The dots stand for more terms as if an expansion could be expected. Eq. (14) is intended as a phenomenological ansatz only; whether it could come from an expansion or not is something we shall not go into. Ansätze such as Eq. (14) would be interesting if they describe experiment much better than other possibilities and if the parameters $A_1, A_2, \ldots$ point towards some simple scheme. Without persuing this scheme further, let us say that at least it is interesting to try to connect the Cabibbo angle with some function of the hyperon masses.

In brief, the SG models we shall consider are a Cabibbo —like one— with the hadronic current given by (9), (10) —and a multicommutator SG—with Cabibbo-universality (9), (10) replaced by Eq. (14).

9.3. Comparison with Experiment.

In order to compare[5] with experiment we know we must incorporate radiative corrections and the q^2-dependence of the form factors. This all we shall do as before. The only difference now is that it is $\hat{q}^2$ that appears in the F_i and G_i form factors. They can be expanded as

$$F_{1,2}^{\alpha'\beta\alpha}(\hat{q}^2) = \sum_{\gamma=1,2} C(\gamma;\alpha'\beta\alpha)\left[F_{1,2}^{\gamma}(0) + \lambda_{F_{1,2}}^{\gamma}\hat{q}^2\right], \tag{15}$$

and

$$G_{1,3}^{\alpha'\beta\alpha}(\hat{q}^2) = \sum_{\gamma=1,2} C(\gamma;\alpha'\beta\alpha)\left[G_{1,3}^{\gamma}(0) + \lambda_{G_{1,3}}^{\gamma}\hat{q}^2\right] \tag{16}$$

The slope parameters $\lambda_{F_{1,2}}^{\gamma}$ are determined by CVC and the q^2-dependence of the nucleon electromagnetic form factors, just as was done in Sec. 6.1. Thus all eight parameters of the V_μ current are fixed by CVC.

The $\hat{q}^2$-dependence of G_1 and G_3 cannot be fully determined for the same reason the q^2-dependence of g_1 was not determined either. But we can again use what is known on $g_1^{np}(q^2)$,

$$g_1^{np}(q^2) \simeq g_1^{np}(0)\left(1 + \frac{2q^2}{M_A^2}\right), \tag{17}$$

as in Sec. 6.1. Since we know that small differences in the slope parameters do not really matter we shall assume a dipole q^2-dependence for all other g_1 form factors.

Eqs. (11) get a correction due to the $\hat{q}^2$- and q^2-dependences. For the g_1 form factors the result is

$$g_1^{\alpha'\beta\alpha}(0) = \sum_{\gamma} C(\gamma,\alpha'\beta\alpha)\left[G_1^{\gamma}(0) + \frac{m_\alpha^2 - m_{\alpha'}^2}{2m_\alpha m_{\alpha'}}G_3^{\gamma}(0)\right]$$

$$\times\left[1 + \frac{2(m_\alpha - m_{\alpha'})^2}{M_A^2}\right]^{-1}. \tag{18}$$

For f_1 an analogous change occurs. There is no need to give it explicitly.

We can now proceed to compare with experiment. Results are displayed in Tables 1,1(A), and 1(B) for the Cabibbo-like SG models, and in Tables 2 and 2(A) for the multicommutator SG models. In Table 1(a) no higher representations are used, and in Table 1(b) higher representations are introduced with the Cabibbo-rotated current of Eq. (8.7). The Cabibbo-like SG model without higher representation is able to explain α_e in $\Sigma^- \to ne\nu$, but then it cannot reproduce the $\Sigma^- \to \Lambda e\nu$ rate. When higher representations are introduced the fit is very good, χ^2 is lowered from 44.2 to 19.6. The contribution of higher representations required to explain the $\Sigma^- \to \Lambda e\nu$ rate along with α_e in $\Sigma^- \to ne\nu$ is rather small as can be seen in Table 1(A). The reduced form factors of 10, $\overline{10}$, and 27 are around 5 to 10% of the leading octet reduced form factors. In this sense the Cabibbo-like SG predicts that the A_μ-current is an SU(3) octet, but it is not a pure one and small admixtures of higher representations are present as well.

Results for the multicommnutator SG models are displayed in Tables 2 and 2(A). Higher representations are introduced using Eq. (8.8). Without higher representations and with only two commutators the multicommutator ansatz has strong discrepancies with the most recent values of the $\Xi^- \to \Lambda e\nu$ and $\Xi^- \to \Sigma^0 e\nu$ rates. It predicts for them the opposite behavior the CT does; $\Xi^- \to \Lambda e\nu$ is predicted to be some 6 times more frequent that $\Xi^- \to \Sigma^0 e\nu$ in the CT, while the multicommutator Ansatz predicts roughly the reversed situation, there is no need to display a table. In Table 2(a) four terms in the multi-commutator expansion are used. The strong discrepancy in $\Xi^- \to \Lambda e\nu$ is corrected, but a discrepancy with $\Sigma^- \to \Lambda e\nu$ appears. They are all corrected if higher representations are introduced. The results when only two commutators are used are displayed in Table 2(b), and when three commutators are used Table 2(c) is obtained. χ^2 is brought from 124.1 down to 31.3 and 26.9, respectively. All the rates are well fit, but some deviations remain in the angular coefficients of $\Sigma^- \to \Lambda e\nu$ and $\Xi^- \to \Lambda e\nu$.

9.4. Discussion. The Best Prediction of Spectrum Generating SU(3).

From the χ^2 point of view, a good result obtained in the last section is the one with the multicommutator ansatz and with higher representations, Table 2(c). However due to the larger number of parameters involved and to the large values of several of them this SG model does not appear to be an optimal solution of the problem of SU(3) corrections

TABLE 1. Comparison of the Cabibbo-like SG model without and with higher representations, columns (a) and (b), respectively. The data of Table 3.4 are used; radiative corrections and q^2-dependence of f_1 and g_1 are accounted for (see text). The values of the parameters are given in Table 1(A), and the explicit values for the form factors for fit 1(b) are given in Table 1(B).

	(a)		(b)	
	Prediction	$\Delta\chi^2$	Prediction	$\Delta\chi^2$
$n \to pe\nu$ (rate)	1.077	0.09	1.084	0.05
$\Sigma^+ \to \Lambda e\nu$ (rate)	0.280	0.59	0.238	0.03
$\Sigma^- \to \Lambda e\nu$ (rate)	0.459	15.49	0.390	0.02
$\Lambda \to pe\nu$ (rate)	3.248	1.17	3.182	0.00
$\Sigma^- \to ne\nu$ (rate)	6.751	0.91	6.904	0.06
$\Xi^- \to \Lambda e\nu$ (rate)	3.017	2.66	3.326	0.01
$\Xi^- \to \Sigma^0 e\nu$ (rate)	0.555	0.06	0.654	1.53
$\Lambda \to p\mu\nu$ (rate)	0.559	0.10	0.553	0.13
$\Sigma^- \to n\mu\nu$ (rate)	3.042	0.00	3.155	0.18
$\Xi^- \to \Lambda\mu\nu$ (rate)	0.821	0.38	0.904	0.33
$n \to pe\nu$ ($\alpha_{e\nu}$)	-0.073	0.04	-0.075	0.04
$n \to pe\nu$ (α_e)	-0.080	1.95	-0.082	0.18
$n \to pe\nu$ (α_ν)	0.989	0.12	0.989	0.13
$\Sigma^\pm \to \Lambda e\nu$ ($\alpha_{e\nu}$)	-0.465	0.59	-0.469	0.63
$\Sigma^- \to \Lambda e\nu$ ($\alpha_{e\nu}$)	-0.475	2.61	-0.480	2.95
$\Sigma^- \to ne\nu$ ($\alpha_{e\nu}$)	0.302	0.75	0.281	0.01
$\Sigma^- \to ne\nu$ (α_e)	0.145	0.37	0.090	0.80
$\Lambda \to pe\nu$ ($\alpha_{e\nu}$)	-0.028	1.16	-0.014	0.01
$\Lambda \to pe\nu$ (α_e)	0.003	3.40	0.008	3.17
$\Lambda \to pe\nu$ (α_ν)	0.973	6.45	0.960	5.40
$\Lambda \to pe\nu$ (α_p)	-0.571	0.95	-0.565	0.76
$\Sigma^- \to \Lambda e\nu$ (A)	0.045	0.13	0.049	0.01
$\Sigma^- \to \Lambda e\nu$ (B)	0.912	0.78	0.914	0.83
$\Xi^- \to \Lambda e\nu$ ($\alpha_{e\nu}$)	0.726	3.83	0.663	1.76
$\Xi^- \to \Lambda e\nu$ (A)	0.625	0.00	0.689	0.48
χ^2		44.21		19.59

TABLE 1(A). Values of the fitted parameters corresponding to fits (a) and (b) of Table 1. See also Table 1(B).

(a)	(b)
$G_1^1 = 1.009 \pm 0.016$	$G_1^1 = 1.040 \pm 0.029$
$G_1^2 = -1.520 \pm 0.013$	$G_1^2 = -1.467 \pm 0.026$
$G_3^1 = -16.0 \pm 3.6$	$G_3^1 = -11.7 \pm 6.9$
$G_3^2 = -42.9 \pm 3.5$	$G_3^2 = -42.5 \pm 4.3$
$\sin\theta = 0.246 \pm 0.002$	$\sin\theta = 0.251 \pm 0.006$
$G_{\overline{10}}$ -	$G_{\overline{10}} = 0.003 \pm 0.035$
G_{10} -	$G_{10} = -0.104 \pm 0.024$
G_{27} -	$G_{27} = -0.073 \pm 0.035$
$\chi^2 = 44.21$	$\chi^2 = 19.59$

TABLE 1(B). For completeness, we give in this table the explicit values of the form factors for the fit of Table 1(b), which is the best prediction of the SG models (see later in the text). The induced form factors displayed in this table are normalized as in Eq. (2.12).

	f_1	g_1	f_2	g_2	f_3	g_3
np	1	1.252	1.854	-0.013	-0.001	18.502
$\Sigma^+\Lambda$	-0.002	0.584	1.210	-0.628	-0.039	19.625
$\Sigma^-\Lambda$	-0.003	0.575	1.214	-0.696	-0.043	19.694
Λp	-1.133	-0.726	-1.202	0.345	0.104	-3.993
$\Sigma^- n$	-0.904	-0.380	1.156	-3.847	-0.139	31.896
$\Xi^-\Lambda$	1.125	0.432	-0.081	1.415	0.007	16.766
$\Xi^-\Sigma^0$	0.677	0.860	1.381	-0.707	-0.071	13.786

TABLE 2. Comparison of multicommutators SG model. The data of Table 3.4 are used. The values of the fitted parameters are given in Table 2(A).

	(a) Prediction	(a) $\Delta\chi^2$	(b) Prediction	(b) $\Delta\chi^2$	(c) Prediction	(c) $\Delta\chi^2$
$n \to pe\nu$ (rate)	1.091	0.56	1.083	0.03	1.083	0.03
$\Sigma^+ \to \Lambda e\nu$ (rate)	0.195	0.77	0.259	0.02	0.259	0.02
$\Sigma^- \to \Lambda e\nu$ (rate)	0.267	45.08	0.387	0.00	0.382	0.12
$\Lambda \to pe\nu$ (rate)	3.149	0.41	3.205	0.11	3.187	0.00
$\Sigma^- \to ne\nu$ (rate)	7.063	0.22	7.005	0.04	7.015	0.06
$\Xi^- \to \Lambda e\nu$ (rate)	3.456	0.66	3.184	0.49	3.310	0.00
$\Xi^- \to \Sigma^o e\nu$ (rate)	1.120	34.84	0.555	0.06	0.582	0.27
$\Lambda \to p\mu\nu$ (rate)	0.560	0.10	0.565	0.07	0.554	0.13
$\Sigma^- \to n\mu\nu$ (rate)	2.603	2.62	2.959	0.09	2.838	0.56
$\Xi^- \to \Lambda\mu\nu$ (rate)	0.940	0.31	0.832	0.37	0.840	0.37
$n \to pe\nu$ ($\alpha_{e\nu}$)	-0.076	0.33	-0.075	0.07	-0.075	0.07
$n \to pe\nu$ (α_e)	-0.084	0.25	-0.082	0.06	-0.082	0.07
$n \to pe\nu$ (α_ν)	0.988	0.15	0.989	0.14	0.989	0.14
$\Sigma^\pm \to \Lambda e\nu$ ($\alpha_{e\nu}$)	-0.539	1.59	-0.496	0.94	-0.501	1.02
$\Sigma^- \to \Lambda e\nu$ ($\alpha_{e\nu}$)	-0.556	11.94	-0.508	5.64	-0.515	6.34
$\Sigma^- \to ne\nu$ ($\alpha_{e\nu}$)	0.271	0.10	0.277	0.01	0.271	0.09
$\Sigma^- \to ne\nu$ (α_e)	0.342	0.19	0.267	0.00	0.310	0.07
$\Lambda \to pe\nu$ ($\alpha_{e\nu}$)	0.015	4.10	-0.004	0.39	-0.009	0.07
$\Lambda \to pe\nu$ (α_e)	-0.028	5.36	-0.022	4.99	-0.008	4.07
$\Lambda \to pe\nu$ (α_ν)	0.863	0.48	0.890	0.13	0.921	2.78
$\Lambda \to pe\nu$ (α_p)	-0.476	0.24	-0.498	0.03	-0.527	0.09
$\Xi^- \to \Lambda e\nu$ ($\alpha_{e\nu}$)	0.786	6.57	0.216	9.87	0.392	1.90
$\Xi^- \to \Lambda e\nu$ (A)	0.377	5.90	0.852	5.38	0.894	7.49
$\Sigma^- \to \Lambda e\nu$ (A)	0.038	0.20	0.040	0.19	0.039	0.20
$\Sigma^- \to \Lambda e\nu$ (B)	0.926	1.19	0.917	0.93	0.918	0.96
χ^2		124.10		31.28		26.90

TABLE 2(A). Values of the fitted parameters of the multicommutators SG model. Columns (a), (b), and (c) match the corresponding ones in Table 2.

G_1^1	0.58 ± 0.05	0.95 ± 0.13	0.83 ± 0.11
G_1^2	-1.87 ± 0.04	-1.22 ± 0.07	-1.30 ± 0.07
G_3^1	-20.8 ± 4.4	-5.4 ± 6.6	-16.5 ± 9.3
G_3^2	-118.5 ± 9.9	-74.9 ± 11.0	-82.4 ± 11.0
$G_{\overline{10}}$	-	0.54 ± 0.13	0.53 ± 0.10
G_{10}	-	-0.47 ± 0.06	-0.36 ± 0.07
G_{27}	-	-0.69 ± 0.06	-0.49 ± 0.10
A_1	$(-2.87 \pm 0.17) \times 5^2$	$(-1.32 \pm 0.02) \times 5^2$	$(-1.57 \pm 0.13) \times 5^2$
A_2	$(4.24 \pm 0.53) \times 5^4$	$(0.52 \pm 0.02) \times 5^4$	$(0.95 \pm 0.21) \times 5^4$
A_3	$(-2.69 \pm 0.52) \times 5^6$	-	$(-0.18 \pm 0.08) \times 5^6$
A_4	$(0.60 \pm 0.16) \times 5^8$	-	-
χ^2	124.10	31.28	26.90

in HSD. We would conclude that the new[3,9] rates of $\Sigma^- \to \Lambda e\nu$ and $\Xi^- \to \Sigma^0 e\nu$, rule out the ansatz of Eq. (14) for the weak-current operator. The SG model with Cabibbo universality explains all experimental data available at present, if small higher-representation corrections are introduced; this is then the best prediction of the SG approach (Table 1(b)).

Our conclusion for a simple ansatz such as Eq. (14) with only a few terms and with a simple form of the A_μ current where the Cabibbo angle is related to symmetry breaking, attractive as it may be, shows clearly that the Cabibbo universality is an assumption strongly supported by HSD. The challenge as to how to relate the Cabibbo angle with physical parameters such as masses and fundamental constants remains open, however.

In Sec. 8.2 we saw that g_2 plays a role in the CT if a positive α_e in $\Sigma^- \to ne\nu$ is to be predicted, but as we discussed there g_2 is required to be so large that the first-order symmetry-breaking limitation of Eqs. (8.3) and (8.4) is violated. From this we concluded that the current value of α_e in $\Sigma^- \to ne\nu$ can only be explained by strong symmetry-breaking. One may ask[5] if a g_2 as suggested by SG is all that is needed in the CT. If a term of the form (an SU(3) octet)

$$g_2^{\alpha'\beta\alpha} = \sum_{\gamma=1,2} C(\gamma,\alpha'\beta\alpha) \frac{m_\alpha - m_{\alpha'}}{2m_\alpha m_{\alpha'}} g_2^\gamma$$

is added to the CT, instead of Eq. (8.4), the prediction for α_e in $\Sigma^- \to ne\nu$ is lowered to -0.57. It is still too far from a positive α_e. It is then not only g_2 that can explain a positive α_e, but the simultaneous change in g_1 that does it. In Sec. 8.2 this fact reflected itself in large parameters for symmetry breaking in g_1. This means then that the full SG assumptions and not only their prediction for g_2 are needed to explain the new world average of α_e in $\Sigma^- \to ne\nu$. The admixtures of higher representations in SG are very reminiscent of the situation in hyperon magnetic moments[6]. The SG model —predicting magnetic moments in intrinsic magnetons— gave a good overall description of experimental data, but in order to explain the precision data higher representations must be added as a small correction to the electromagnetic current operator.

Among the HSD observables α_e in $\Sigma^- \rightarrow ne\nu$ has a unique position. The predictions for it by the CT and SG approach are so far apart from one another that small perturbative-like effects —such as small SU(3) symmetry breaking, radiative corrections, q^2-dependence of form factors, higher representations— cannot mask their difference. This is a most fortunate situation, since low-statistics experiments that reduce the error bars on this α_e from 0.19 to something between 0.05 and 0.10 can easily decide whether strong symmetry-breaking is present in HSD. If this were to be the case in the future, one would then be forced to accept that a non-perturbative approach to deal with SU(3) symmetry breaking is required and that establishing the set of commuting operators from the outset should be a must. Approaches such as the SG one could then be most useful.

Chapter 10. High Statistics Experiments

10.1. Analysis of Dalitz Plots. Some Theorems.

We shall now abandon the theoretical predictions for HSD and we shall make a complete turn into some simple results which can be very useful in experimental analysis. So far we have discussed the analysis of low statistics experiments —with hundreds or several thousands of events. We now come[1] to consider how the analysis of experiments with tens of thousands of events and, even more, hundreds of thousands of events can be made. Currently, experiments with up to 10^5 events have been partly analyzed. And, the present state-of-the-art of experimental techniques allows already that several million events of some HSD be collected.

At the beginning in Chapter 3, we made it clear that low statistics experiments could not be analyzed in terms of Dalitz plots —the differential decay rate— and information had to be collected into integrated observables. The frequent use of quoting experimental results in terms of g_1/f_1 ratios, implies that some kind of detailed analysis has been performed, as if a Dalitz plot were being measured. In this way, the common belief that form factors are observable, or directly accesible to experimental analysis, is upheld. We have stressed several times that this is not the case. To have access to a g_1/f_1 ratio one must have previously determined the other form factor ratios —two more in e-mode decays and four more in μ-mode decays. This is always done not by measuring the other ratios —although attempts are sometimes made— but by fixing them theoretically. This practice renders the so-measured g_1/f_1 ratios model dependent. The values obtained are useful in as much as one compares them with other values —theoretical or experimental— that use the same extra assumptions. Thus, the experimental value of a g_1/f_1 ratio obtained under the assumption that there is no symmetry breaking cannot be used for comparisons with predictions for it that assume symmetry breaking effects. Clearly, the use of experimental evidence on g_1/f_1 ratios is limited to rough comparisons only. This has been our major objection to use g_1/f_1 ratios. But the question remains: to what extent are form factors and form-factor ratios measurable? As we shall see in this chapter, the answer to this question is subtle. It is not the form factors that are easily observable; instead, quadratic combinations of them are. This can be seen by looking at the kinematical variables that, after all, identify events. This fact has other consequences as well, it occurs that no matter how precisely a certain Dalitz plot is measured one may still not be able to uniquely determine all the form factors. For example, it is often believed that the energy spectrum of the emitted electron can determine the sign of g_1/f_1, even if the rate and the $\alpha_{e\nu}$ correlation coefficient could not. This is not the case.

In what follows we shall limit ourselves to e-mode HSD, in order to keep expressions manageable, but our results can be straight-forwardly extended to μ-modes. The long and tedious expressions for the differential decay rate including radiative corrections of Eqs. (2.17)-(2.20) and (2.25)-(2.26) and (5.15)-(5.20) can be turned into simple looking expressions. We shall state this in the form of three theorems and an approximate form of each one of them.

Theorem 1. Within the V-A theory and neglecting the mass of the emitted electron or positron (wherever there are no mass singularities), the Dalitz plot of unpolarized decaying hyperons (UPDP) can be expressed as a linear combination of six functions of q^2, namely,

$$D_1' = A(q^2) + \frac{E}{M_1} C(q^2) + \frac{E_\nu}{M_1} E'(q^2), \tag{1}$$

$$D_2' = B(q^2) + \frac{E}{M_1} D(q^2) + \frac{E_\nu}{M_1} F(q^2). \tag{2}$$

The proof of this theorem is simple and we shall not give it. The six functions are defined —in terms of the form factors of Eqs. (2.17)-(2.18)— as

$$\begin{aligned} A(q^2) = {} & (2 - M_2/M_1)F_1^2 + \frac{1}{2}(1 + M_2/M_1)F_2^2 \\ & + (1 + M_2/M_1)F_1F_2 + (2 + M_2/M_1)G_1^2 \\ & + \frac{1}{2}(1 - M_2/M_1)G_2^2 + (-1 + M_2/M_1)G_1G_2, \end{aligned}$$

$$B(q^2) = (M_2/M_1)F_1^2 + \frac{1}{2}(1+M_2/M_1)F_2^2$$

$$+ (1+M_2/M_1)F_1F_2 - (M_2/M_1)G_1^2$$

$$+\frac{1}{2}(1-M_2/M_1)G_2^2 + (-1+M_2/M_1)G_1G_2,$$

$$C(q^2) = -F_1^2 - \frac{1}{2}F_2^2 - G_1^2 - \frac{1}{2}G_2^2 + 2F_1G_1,$$

$$D(q^2) = F_1^2 - \frac{1}{2}F_2^2 + G_1^2 - \frac{1}{2}G_2^2 - 2F_1G_1,$$

$$E'(q^2) = -F_1^2 - \frac{1}{2}F_2^2 - G_1^2 - \frac{1}{2}G_2^2 - 2F_1G_1,$$

$$F(q^2) = F_1^2 - \frac{1}{2}F_2^2 + G_1^2 - \frac{1}{2}G_2^2 + 2F_1G_1. \tag{3}$$

The primes on D_1' and D_2' are a reminder that radiative corrections have been taken into account. We follow the same conventions as in Chapter 5. In particular, the effective form factors of Eqs. (5.11) and (5.12) should be understood to appear in Eqs. (3).

When the initial hyperon is polarized, we get for the Dalitz plot that contains the information on this polarization (we call this Dalitz plot PDP)[3]:

<u>Theorem 2</u>. Within the V-A theory and neglecting the mass of the emitted electron or positron (wherever no mass singularities appear), the PDP can be expressed as a linear combination of seven functions of q^2, namely,

$$D_3' = A_1 + \frac{E}{M_1}C_1 + \frac{E_\nu}{M_1}E_1 + x_eG_1', \tag{4}$$

$$D_4' = B_1 + \frac{E}{M_1}D_1 + \frac{E_\nu}{M_1}F_1' + x_\nu H_1, \tag{5}$$

(only the combination $D_1 + G_1'$ is kinematically accesible).

Again we shall omit the proof of this theorem. We use the same definitions of Eqs. (2.19) and (2.20) with radiative corrections as in Eq. (5.15) understood. The eight functions of Eqs. (4) and (5) are

$$A_1 = -(1-M_2/M_1)F_1^2 - (1+M_2/M_1)G_1^2 + 2F_1G_1$$
$$- (1-M_2/M_1)F_1G_2 + (1+M_2/M_1)G_1F_2,$$
$$B_1 = (1-M_2/M_1)F_1^2 + (1+M_2/M_1)G_1^2 + 2F_1G_1$$
$$- (1-M_2/M_1)F_1G_2 + (1+M_2/M_1)G_1F_2,$$
$$C_1 = F_1^2 + G_1^2 - 2F_1G_1 - F_2G_2,$$
$$D_1 = F_1F_2 - G_1G_2 + F_1G_2 - G_1F_2,$$
$$E_1 = -F_1F_2 + G_1G_2 + F_1G_2 - G_1F_2,$$
$$F_1' = -F_1^2 - G_1^2 - 2F_1G_1 - F_2G_2,$$
$$G_1' = -F_1^2 - G_1^2 + 2F_1G_1 - F_1F_2 + G_1G_2 - F_1G_2$$
$$+ G_1F_2 - F_2G_2,$$
$$H_1 = F_1^2 + G_1^2 + 2F_1G_1 + F_1F_2 - G_1G_2 - F_1G_2$$
$$+ G_1F_2 - F_2G_2. \tag{6}$$

The third theorem is the equivalent of Theorem 2 for the case when it is the polarization of the emitted hyperon that is observed. We call this Dalitz plot P'DP. Thus[3],

Theorem 3. Within the V-A theory and neglecting the mass of the emitted electron or positron (whenever no mass singularities appear), the P'DP can be expressed as a linear combination of seven functions of $\hat{q}^2$, namely,

$$\hat{D}_5' = A_2(\hat{q}^2) + \frac{\hat{E}}{M_1} C_2(q^2) + \frac{\hat{E}_\nu}{M_1} E_2(\hat{q}^2) + \hat{x}\frac{\hat{E}}{M_1} G_2'(\hat{q}^2), \tag{7}$$

$$\hat{D}_6' = B_2(\hat{q}^2) + \frac{\hat{E}}{M_1} D_2(\hat{q}^2) + \frac{\hat{E}_\nu}{M_1} F_2'(\hat{q}^2) + \hat{x}\frac{\hat{E}_\nu}{M_1} H_2(\hat{q}^2). \tag{8}$$

(Only the combination $D_2 + G_2'$ is kinematically accesible).

The proof is also omitted. The P'DP is given in the center of mass of the emitted hyperon and all conventions are as in Eqs. (2.25), (2.26) and (5.18). The eight functions of Eqs. (7) and (8) are

$$A_2 = (1 - M_1/M_2)F_1^2 + (1 + M_1/M_2)G_1^2 + 2F_1G_1$$

$$+ (-1 + M_2/M_1)F_1G_2 + (1 + M_2/M_1)G_1F_2,$$

$$B_2 = (-1 + M_1/M_2)F_1^2 - (1 + M_1/M_2)G_1^2 + 2F_1G_1$$

$$+ (-1 + M_2/M_1)F_1G_2 + (1 + M_2/M_1)G_1F_2,$$

$$C_2 = (M_1/M_2)(F_1^2 + G_1^2 + 2F_1G_1)$$

$$- (M_2/M_1)F_2G_2,$$

$$D_2 = F_1F_2 + G_1G_2 + F_1G_2 + G_1F_2,$$

$$E_2 = -F_1F_2 - G_1G_2 + F_1G_2 + G_1F_2,$$

$$F_2' = -(M_1/M_2)(F_1^2 + G_1^2 - 2F_1G_1)$$

$$- (M_2/M_1)F_2G_2,$$

$$G_2' = -(M_1/M_2)(F_1^2 + G_1^2 + 2F_1G_1) - F_1F_2 - G_1G_2$$

$$- F_1G_2 - G_1F_2 - (M_2/M_1)F_2G_2,$$

$$H_2 = (M_1/M_2)(F_1^2 + G_1^2 - 2F_1G_1) + F_1F_2 + G_1G_2$$

$$- F_1G_2 - G_1F_2 - (M_2/M_1)F_2G_2. \qquad (9)$$

The q^2 or $\hat{q}^2$ of the different Dalitz plots comes from two places. One is from E_ν and $\hat{E}_\nu$ and the other one is the form factor q^2-dependence. The former is given by the relationship

$$\frac{E_\nu}{M_1} = \frac{M_1^2 - M_2^2}{2M_1^2} - \frac{E}{M_1} + \frac{q^2}{2M_1^2}$$

and an anlogous one for $\hat{E}_\nu$. The latter can be approximated by expanding the leading form factors as in Eqs. (3.11) and (3.12), but we shall use the following conventions now

$$f_1(q^2) = f_1(0)\left(1 + \lambda_{f_1} \frac{q^2}{M_1^2}\right) ,$$

$$g_1(q^2) = g_1(0)\left(1 + \lambda_{g_1} \frac{q^2}{M_1^2}\right) ,$$

because they lead to simpler equations.

Using these two expansions, the three theorems become then, within the approximations of the above expansions[3],

Theorem 1'. Within the V-A theory ... the UPDP ... in terms of six constants ...

$$D_1' = A' + \frac{E}{M_1} C' + \frac{q^2}{M_1^2} E'' , \tag{10}$$

$$D_2' = B' + \frac{E}{M_1} D' + \frac{q^2}{M_1^2} F' . \tag{11}$$

Theorem 2'. Within ... the PDP ... in terms of seven constants ...

$$D_3' = A_1' + \frac{E}{M_1} C_1' + \frac{q^2}{M_1^2} E_1' + x_e G_1' , \tag{12}$$

$$D_4' = B_1' + \frac{E}{M_1} D_1' + \frac{q^2}{M_1^2} F_1'' + x_\nu H_1 . \tag{13}$$

Theorem 3'. Within ... the P'DP ... in terms of seven constants ...

$$\hat{D}_5' = A_2' + \frac{\hat{E}}{M_1} C_2' + \frac{\hat{q}^2}{M_1^2} E_2' + \hat{x} \frac{\hat{E}}{M_1} G_2' \tag{14}$$

$$\hat{D}_6' = B_2' + \frac{\hat{E}}{M_1} D_2' + \frac{\hat{q}^2}{M_1^2} F_2'' + \hat{x} \frac{\hat{E}_\nu}{M_1} H_2 . \tag{15}$$

In the last equation we have kept $\hat{E}_\nu$ because of the factor $\hat{x}$ in front of it, just as x_ν is kept in D_4'.

The constants are ($\hat{E}_m = E_m$ if m is dropped)

$$A' = A + (E_m/M_1)E',$$

$$B' = B + (E_m/M_1)F,$$

$$C' = C - E',$$

$$D' = D - F,$$

$$E'' = \frac{1}{2} E' + \lambda,$$

$$F' = \frac{1}{2} F + \lambda', \tag{16}$$

$$A_1' = A_1 + (E_m/M_1)E_1,$$

$$B_1' = B_1 + (E_m/M_1)F_1',$$

$$C_1' = C_1 - E_1,$$

$$D_1' = D_1 - F_1',$$

$$E_1' = \frac{1}{2} E_1 + \lambda_1,$$

$$F_1'' = \frac{1}{2} F_1' + \lambda_1', \tag{17}$$

$$A_2' = A_2 + (E_m/M_1)E_2,$$

$$B_2' = B_2 + (E_m/M_1)F_2',$$

$$C_2' = C_2 - E_2,$$

$$D_2' = C_2 - F_2',$$

$$E_2' = \frac{1}{2} (M_1/M_2)E_2 + \lambda_1'$$

$$F_2'' = - \frac{1}{2} (M_1/M_2)F_2' + \lambda_1. \tag{18}$$

λ, λ', λ_1, and λ_1' are

$$\lambda = 2(F_1 + F_2)^2 \lambda_{f_1} + 6G_1^2 \lambda_{g_1}, \tag{19}$$

$$\lambda' = 2(F_1 + F_2)^2 \lambda_{f_1} - 2G_1^2 \lambda_{g_1}, \tag{20}$$

$$\lambda_1 = -4G_1^2 \lambda_{g_1} + 2G_1(F_1 + F_2)(\lambda_{f_1} + \lambda_{g_1}), \tag{21}$$

$$\lambda_1' = 4G_1^2 \lambda_{g_1} + 2G_1(F_1 + F_2)(\lambda_{f_1} + \lambda_{g_1}). \tag{22}$$

The expressions for G_1', H_1, G_2', and H_2 remain unchanged from the corresponding ones in Eqs. (6) and (9). It should be understood that now $A' = A'(0)$, etc. Throughout Eqs.

(16)-(18). The validity of the "primed" theorems is within a percent, so they are quite precise.

The above theorems are very simple, but as we shall see in the next section, they have non-trivial consequences. It should be very clear now that it is the constants defined in Eqs. (16)-(18) that are easily accesible to experiment, instead of the form factor ratios themselves.

10.2. Determination of Form Factors. Tests of the V-A Theory.

No matter how the analysis of a Dalitz plot of HSD is performed one has no choice but to pass through the constants of Eqs. (16)-(18) or their equivalents if another set of kinematical variables is selected. It is then these quantities that can be directly quoted in an experiment. Only as a second step can form factors be measured, once the values of those constants has been determined. One may not do this explicitly, but one is still going through a two step procedure anyway.

Assuming that those constants are known, we can proceed to extract information on the form factors from them. We shall first consider the case where no hyperon polarization is available, i.e., the UPDP case. The first consequence of Theorem 1 is that, although there are six functions at our disposal, Eqs. (3), they are not all independent. One can readily see that

$$C(q^2) - E'(q^2) = -D(q^2) + F(q^2). \tag{23}$$

This equality is a consequence of the V-A structure of Eqs. (1) and (2). If experimentally Eq. (23) is not fulfilled then it should be concluded that the V-A theory is not strictly valid. It is important to remark that this test is locally valid, i.e., for each pair of values of E and x. In the case of Theorem 1', Eq. (23) becomes

$$C' = -D' \tag{24}$$

All along in the past chapters we have assumed the validity of the V-A theory. In a high statistics experiment one can perform certain tests to see if the V-A theory fails. This feature is gratifying but it has its price. Immediately, we can conclude that no matter how precisely the UPDP of HSD is measured it is impossible to uniquely determine all the relevant form factors. They are four functions of q^2 and the first approximation to them cannot be determined, because there are six unknowns — four form factors at $q^2 = 0$ and two slopes — and only five independent constants to be experimentally measured. Therefore, no unique solution for all the form factors exists in the UPDP of HSD.

In order to establish which form factors can be uniquely determined, we can form the following equations:

$$A' - B' = \left(\frac{M_1-M_2}{M_1} F_1 - \frac{M_1+M_2}{M_1} G_1 \right)^2 , \tag{25}$$

$$C' - D' = 8F_1G_1 , \tag{26}$$

$$E'' - F' = -(F_1 + G_1)^2 + \lambda - \lambda' , \tag{27}$$

$$A' + B' = 2 \left(F_1 + \frac{M_1+M_2}{2M_1} F_2 \right)^2 + 2 \left(G_1 - \frac{M_1-M_2}{2M_1} G_2 \right)^2 , \tag{28}$$

$$E'' + F' = - \frac{1}{2} (F_2^2 + G_2^2) + \lambda + \lambda' \tag{29}$$

Eqs. (25) and (26) depend only on F_1 and G_1. The first one represents two parallel lines and the second one a hyperbola. Their intersections can give at most four solutions, but since the constants of Eqs. (16) are quadratic combinations with linear terms absent the overall sign of the four factors is always undetermined and can be fixed arbitrarily. Thus, there are only two relevant solutions for F_1 and G_1. Eq. (27) determines the combination λ-λ' in terms of F_1 and G_1. This combination depends only on λ_{g_1}, namely,

$$\lambda - \lambda' = 8G_1^2 \lambda_{g_1} . \tag{30}$$

There are then two solutions allowed for λ_{g_1}, too.

The two remaining equations are an ellipse, Eq. (28), and a hyperbola, Eq. (29), in the F_2, G_2 plane. Their intersections can be four at most. Actually, the role of Eq. (28) is to provide bounds on F_2 and G_2,

$$F_2 \in \frac{2M_1}{M_1+M_2}\left[-F_1-\alpha,-F_1+\alpha\right], \tag{31}$$

$$G_2 \in \frac{2M_1}{M_1-M_2}\left[G_1-\alpha,G_1+\alpha\right], \tag{32}$$

where

$$\alpha = 2^{-1/2}(A'+B')^{1/2}.$$

So, in principle there are two solutions for F_1, G_1, and λ_{g_1}, two sets of bounds on F_2 and G_2, and a correlation between the latter two and λ_{f_1}.

We can say something more. The fact that the mass difference M_1-M_2 appears in front of F_1 in Eq. (25) makes that one of the solutions for F_1 and G_1 has a G_1 of the order of $(1-M_2/M_1)^{-2}$, as can be seen by explicit calculation. For this type of solution Eq. (30) then implies that λ_{g_1} is of the order $(1-M_2/M_1)^{-4}$. Given the hyperon mass differences, λ_{g_1} turns out to close to 10^3, or three orders of magnitude larger than what is expected from analyticity of strong interactions. It must thus be rejected (actually, all of our approximations break up for this sort of λ_{g_1}). Therefore in practice, there exists a unique solution for F_1, G_1, and λ_{g_1}.

The bounds of Eq. (31) on F_2 are somewhat restrictive. α is about the order of magnitude of F_1, as can be seen by looking through Eqs. (3) and (16). Thus F_2 is bounded within a factor of 2 or 3 around F_1. In contrast, the bounds of Eq. (32) on G_2 are very lax, because M_1-M_2 appears in the denominator. One can see that the ellipse of Eq. (28) is very elongated parallel to the G_2-axis; it is approximately some 10 times longer than wider. The four possible intersections (i.e. four sets of F_2, G_2 and λ_{f_1}) of this ellipse and the hyperbola provide essentially a very strong correlation between G_2 and λ_{f_1}. A small change in λ_{f_1} makes the intersections of the hyperbola of Eq. (29) and the ellipse to move very fast along the G_2 direction. We have drawn Eqs. (25), (26), and (28), (29) in Figs. 1 and 2, respectively, to help the reader visualize what has been said above; the form factors predicted for $\Lambda \to pe\nu$ by the CT were used.

To extract information from PDP we can follow the steps analogous to those of the analysis of the UPDP. There is another test of the V-A theory, because of the relation

$$D_1(q^2) + G_1'(q^2) - H_1(q^2) = -C_1(q^2) + F_1'(q^2) + E_1(q^2). \tag{33}$$

Again it is a local test. When the slopes of the form factors are introduced Eq. (33) becomes

$$D_1' + G_1' = H_1 - C_1' \tag{34}$$

Only six out of the seven constants are independent of one another. Combinations analogous to Eqs. (25)-(29) can be formed:

$$A_1' - B_1' + \frac{E_m}{M_1} H_1 = -\left(\frac{M_1 - M_2}{M_1} F_1 - \frac{M_1 + M_2}{M_1} G_1 \right)^2 , \tag{35}$$

$$G_1' + D_1' = 4F_1G_1 , \tag{36}$$

$$E_1' - F_1'' + \frac{1}{2} H_1 = (F_1 + G_1)^2 - 8G_1^2 \lambda_{g_1} , \tag{37}$$

$$A_1' = -\frac{M_1 - M_2}{M_1} F_1(F_1 + G_2) - \frac{M_1 + M_2}{M_1} G_1(G_1 - F_2) + 2F_1G_1 + \frac{M_1^2 - M_2^2}{2M_1}(F_1 + G_1)(-F_2 + G_2) \tag{38}$$

$$B_1' - \frac{E_m}{M_1} C_1' = \frac{M_1 - M_2}{M_1} F_1(F_1 - G_2) + \frac{M_1 + M_2}{M_1} G_1(G_1 + F_2) + 2F_1G_1 + \frac{M_1^2 - M_2^2}{2M_1}\left[(F_1 + G_1)(G_2 - F_2) - 2F_1^2 - 2G_1^2 \right] \tag{39}$$

Comparing Eqs. (35)-(37) with Eqs. (25)-(27) we see that their right hand sides coincide, correspondingly. Therefore, just as before, in practice a unique solution exists for F_1, G_1 and λ_{g_1}. What is different now is the equations for F_2 and G_2. Eqs. (38) and (39) are two straight lines on the F_2-G_2 plane. Their intersection determines

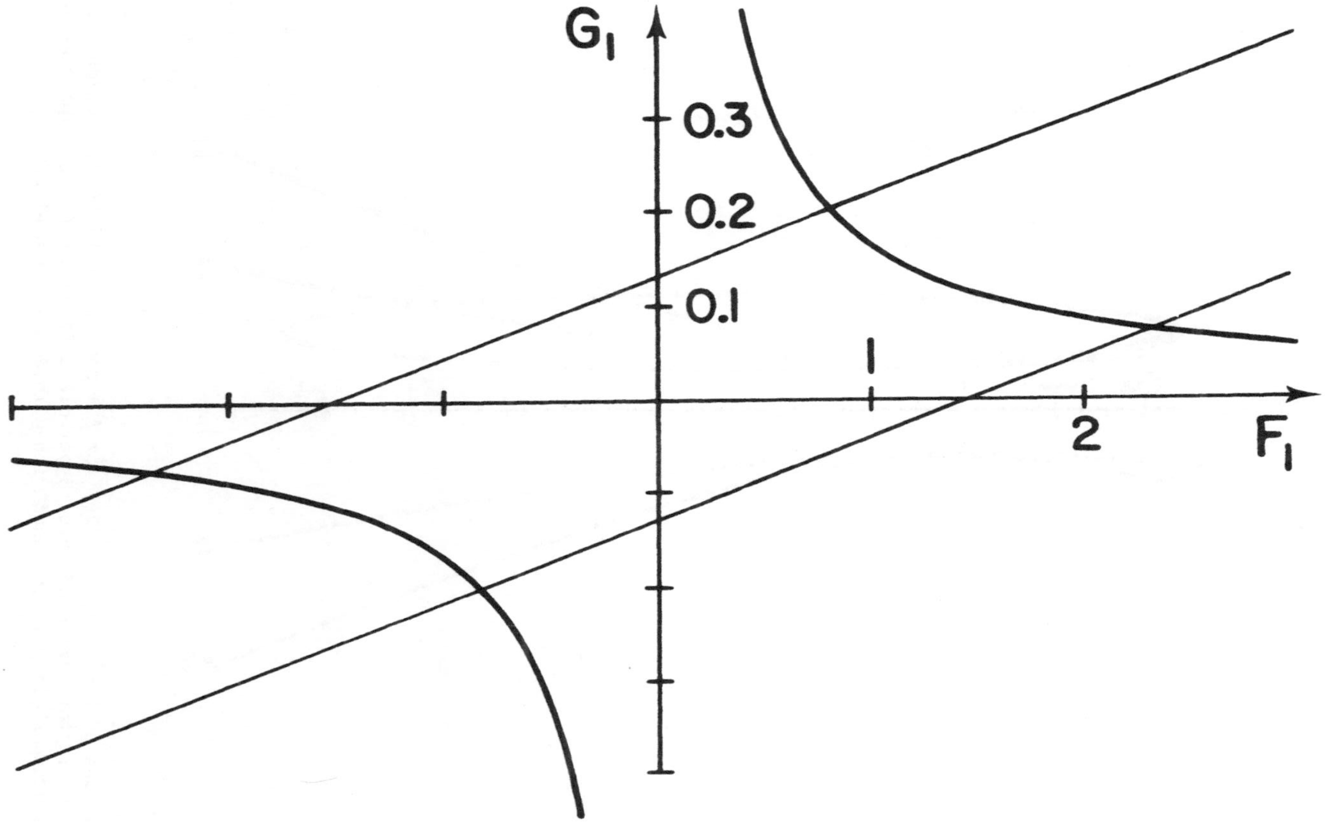

FIG. 1. Graphs of Eqs. (25) and (26) for $\Lambda \to pe\nu$. The Cabibbo predictions for the form factors are used here.

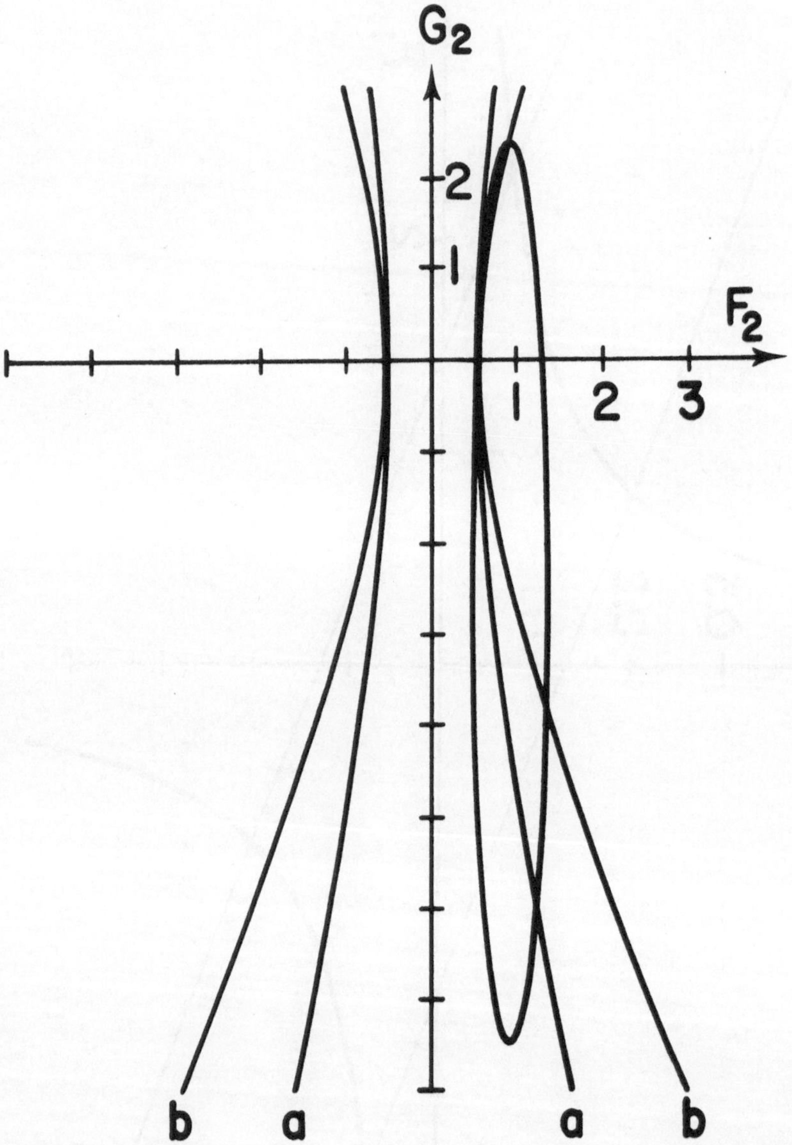

FIG. 2. Graphs of Eqs. (28) and (29) for $\Lambda \to pe\nu$. The Cabibbo predictions for the form factors are used here. Hyperbola a is obtained with dipole A-slopes, and hyperbola b is obtained with pole dominance A-slopes.

a unique solution for F_2 and G_2. λ_{f_1} is determined by one the remaining equations in Eqs. (17), E_1' or F_1''.

The analysis of the P'DP, when the polarization of the emitted hyperon is observed, follows the same pattern as the analysis of PDP. We shall not repeat it here. The interested reader can find the relevant formulas in Ref. (1). The solution is again unique. There is a test of the V-A theory analogous to Eqs. (33) and (34).

Since the values of F_1 and G_1 must be the same in the three Dalitz plots, another two test of the V-A theory are obtained, namely,

$$A_1' - B_1' + \frac{E_m}{M_1} H_1 = -A_2' + B_2' - \frac{E_m}{M_1} H_2 = -A' + B' .$$

Several other tests of the V-A theory can be obtained, in addition to the ones already mentioned. We shall not go any further into this.

To close this section, we want to emphasize that Theorems 1, 2, and 3, and their primed counter-parts show that the experimental evidence of high statistics experiments of HSD can be analyzed in two complementary ways: in terms of the constants of Eqs. (16)-(18) and in terms of form factors directly. The advantage of using the former constants is that their experimental values depend solely on the V-A theory, and in addition the validity of this theory can be tested.

10.3. A "Real Good Experiment".

Besides the general results about the measurability of form factors in HSD of the last two sections, it is convenient to perform a numerical Monte Carlo study to better appreciate the practical usefulness of the proposed parametrization of Dalitz plots in terms of the constants of Eqs. (16), (17), and (18). This study will also give us an estimate about the number of events a high statistics experiment should have.

For definiteness, we shall study $\Sigma^- \to ne\nu$. This processe is particularly suitable because its ΔM is one of the highest available in HSD, and this requires that all the constants of Eqs. (16) and (17) be taken into account. For purposes of illustration, it will suffice that we limit ourselves to consider the CT predictions for this decay. That is, we shall use the corresponding CT form factors and slopes (with dipole λ_{f_1} and

λ_{g_1}) as input to generate with the Monte Carlo method samples with different number of events.

In terms of the variables E and x of Eq. (5.15), the Dalitz plots of $\Sigma^- \to ne\nu$ are simple rectangles. We divide them into 400 bins by taking 20 steps in the range of E and 20 steps in the range of x. Once a sample with a certain number of events has been generated we use it to determine the constants of Eqs. (16), if an unpolarized Σ^- is considered, and the constants of Eqs. (17), if a polarized Σ^- is considered. Finally, we use the fitted values of these constants to obtain a "measurement" of the form factors and slopes, and then we compare the results with their input values.

Our analysis is summarized in Tables 1, 2, and 3. We have considered 5 cases with the number of events N ranging from 10^4 to a little over 3×10^5. The quoted error bars are the full errors obtained by calculating the error matrix. The constants of the UPDP are reproduced in Table 1. As expected, the best determined ones are A' and B', whose error bars are six to eight times smaller than those of C' and D'. The constants E" and F' get the largest error bars, which are about a factor of 50 bigger than those of A' and B'. The relative size of all these error bars is determined by the relative kinematical suppresion of the corresponding constant over the UPDP. The test of the V-A theory through the equality C' = -D' is not very meaningful when $N = 10^4$, it is at the level of 10% when $N = 10^5$, and it begins to be precise when $N \simeq 3 \times 10^5$. All these remarks apply just about as well for the constants of the PDP in Table 2, except that the error bars are typically a factor of 2 compared to those of the constants of the UPDP.

In Table 3, the form factors and slopes are fitted to the constants of Tables 1 and 2 in 5 cases, each with a given number of UPDP events (indicated by N_U) and a given number of PDP events (indicated by N_P), the values obtained should be compared to the input values used. The order of magnitude of G_2 can be seen already at the level of 10^5 unpolarized events if some 10^4 events with good observation of the polarization of Σ^- are available also. It is only in the case of the last column when a refined determination of the form factors is obtained. In this case F_1, F_2, and λ_{f_1} are very precisely "measured" —as was to be expected from our discussion of Sec. 2—, and therefore a very good test of CVC can be performed; G_2 can be bounded around 20% of F_2, and it

TABLE 1. Results for the UPDP constants of $\Sigma^- \to ne\nu$ obtained by fitting them to Monte Carlo generated samples of events. The CT predictions for the form factors of $\Sigma^- \to ne\nu$ were used. The Input column contains the values of the constants that were used to generate the events. N is the number of events that were generated.

	Input	N = 10,000	N = 50,000	N = 100,000	N = 150,000	N = 330,000
A'	0.0613	0.0517 ± 0.0026	0.0602 ± 0.0012	0.0612 ± 0.0009	0.0605 ± 0.0007	0.0609 ± 0.0005
B'	0.0553	0.0562 ± 0.0035	0.0551 ± 0.0015	0.0540 ± 0.0011	0.0546 ± 0.0009	0.0553 ± 0.0006
C'	0.096	0.111 ± 0.019	0.098 ± 0.009	0.104 ± 0.006	0.099 ± 0.005	0.100 ± 0.003
D'	-0.096	-0.051 ± 0.029	-0.081 ± 0.013	-0.084 ± 0.010	-0.087 ± 0.008	-0.095 ± 0.005
E"	0.130	0.400 ± 0.131	0.107 ± 0.060	0.050 ± 0.043	0.144 ± 0.035	0.129 ± 0.094
F'	0.369	0.283 ± 0.106	0.259 ± 0.048	0.297 ± 0.034	0.370 ± 0.028	0.362 ± 0.019
χ^2		381.2	444.5	396.9	425.5	252.2

TABLE 2. Results for the PDP constants of $\Sigma^- \rightarrow ne\nu$ obtained by fitting them to Monte Carlo generated samples of events. Full polarization of Σ^- is assumed. Other conventions are as in Table 1.

	Input	N = 10,000	N = 50,000	N = 100,000	N = 150,000	N = 330,000
A'_1	-0.0240	-0.0207 ± 0.0090	-0.0232 ± 0.0042	-0.0278 ± 0.0030	-0.0241 ± 0.0025	-0.0223 ± 0.0018
B'_1	-0.0333	-0.0368 ± 0.0068	-0.0349 ± 0.0031	-0.0307 ± 0.0022	-0.0327 ± 0.0019	-0.0319 ± 0.0014
C'_1	-0.176	-0.186 ± 0.045	-0.182 ± 0.021	-0.159 ± 0.015	-0.163 ± 0.012	-0.182 ± 0.009
E'_1	-0.017	-0.053 ± 0.384	0.065 ± 0.179	0.154 ± 0.127	-0.025 ± 0.103	-0.085 ± 0.077
F''_1	-0.295	-0.252 ± 0.474	-0.330 ± 0.214	-0.166 ± 0.152	-0.323 ± 0.124	-0.424 ± 0.093
H'_1	-0.080	-0.106 ± 0.080	-0.116 ± 0.036	-0.065 ± 0.026	-0.097 ± 0.021	-0.100 ± 0.016
$D'_1 + G'_1$	0.096	0.142 ± 0.067	0.140 ± 0.030	0.088 ± 0.021	0.101 ± 0.017	0.092 ± 0.013
χ^2		344.1	414.3	330.6	340.4	296.4

TABLE 3. Values of the form factors and slopes obtained by fitting them to the "measured" constants of UPDP and PDP for certain samples of Tables 1 and 2 for $\Sigma^- \to ne\nu$. N_U is the number of events in the UPDP, N_P is the number of events in the PDP used simultaneously with N_U. The asterisk means that the corresponding quantity was kept fixed at its input value.

	Input					
N_U		100	100	330	330	330
N_P		0	10	0	10	50
F_1	0.300	0.302 ± 0.002	0.299 ± 0.016	0.301 ± 0.001	0.311 ± 0.009	0.311 ± 0.009
F_2	-0.591	*	-0.584 ± 0.020	*	-0.601 ± 0.013	-0.602 ± 0.011
λ_{f_1}	-0.69	*	-0.56 ± 0.06	*	-0.68 ± 0.03	-0.68 ± 0.03
G_1	0.080	0.083 ± 0.003	0.083 ± 0.003	0.081 ± 0.002	0.079 ± 0.02	0.079 ± 0.002
G_2	0	*	-0.068 ± 0.133	*	-0.062 ± 0.127	-0.037 ± 0.058
λ_{g_1}	0.15	0.06 ± 0.09	0.07 ± 0.09	0.15 ± 0.05	0.18 ± 0.06	0.18 ± 0.05
χ^2/n_D		4.6/3	3.8/7	2.1/3	1.3/7	3.3/7

would be possible to distinguish between pole-dominance or dipole behavior of λ_{g_1}.

10.4. Remarks.

From the foregoing results we can conclude what a "real good experiment" in HSD should be. First of all it is clear that the number of events must be of the order of several hundreds of thousands of events; at least around a quarter of a million, with about 20% of the events with full polarization of one of the hyperons. Second, the problem of the acceptance of the apparata at this level of statistics should be very carefully considered. In this respect, the lines set by the WA2 collaboration are most welcome. An experiment including several decay modes has a much better control on the systematic biases than several independent experiments each including only one decay mode. At this stage, it becomes evident that a very good high statistics experiment would be one in the style of the WA2-collaboration[3.9], and with several hundreds of thousands of events in each of 5 or 6 decays modes, along with a good hold on the polarization of one the hyperons involved in each decay mode.

The analysis of such an experiment should be performed without fixing any of the form factors nor their relevant q^2-dependence at prejudiced values. Instead, all of the unknowns, so to speak, should be determined from the experiment. The results could be quoted directly through the values of the form factors and their slopes, and through the constants of the UPDP, PDP, and P'DP of the past three sections.

Let us finally remark that the parametrization of the Dalitz plots we are proposing can be further refined by keeping the contributions of the squares of λ_{f_1} and λ_{g_1}; this would improve the precision of the determination of the form factors from the different sets of constants. We have not included their contributions into our analysis; their inclusion is straight forward. At any rate, it should be kept in mind that if f_1 and g_1 are to be accurately determined below 1%, in decays with large ΔM, the incorporation of the q^2-dependence of f_2 becomes necessary.

Chapter 11. Conclusion

We come now to make an evaluation of the material spread over the last 10 chapters. The current experimental situation of hyperon semileptonic decays is far from satisfactory. Out of 16 e-mode decays and 10 μ-mode ones energetically allowed within the hyperon octet only 7 e-mode and 3 μ-mode have been observed and only 3 e-mode decays have their angular coefficients measured with some detail; namely, $n \to pe\nu$, $\Lambda \to pe\nu$, and $\Sigma^- \to ne\nu$. Of the decays that are expected to be absent because of selection rules, only one has an upper bound which could be considered meaningful, $\Sigma^+ \to ne\nu$; the other upper bounds are just too lax still. The efforts of many experimental groups over more than two decades can be poured into a table with only 25 entries. The experimental evidence on hyperon semileptonic decays is scarce. Nevertheless, it already puts constraints on our theoretical beliefs.

The main theory to describe these decays, the Cabibbo theory, has several limitations. As originally stated it has 3 free parameters and is not committed to any particular symmetry-breaking scheme. Nor, does it give an indication about how to incorporate corrections that the precision of the data may require, such as radiative corrections, q^2-dependence of form factors and induced pseudo-scalar form factors g_3 for μ-mode decays. Because of all this the Cabibbo theory was never intended to be exact and discrepancies with experiments should appear eventually. In order to see them one must be sure that the predictive power of the Cabibbo theory is well established. Throughout Chapters 4 and 7 we have studied these questions. Current data require that the working assumptions in the Cabibbo theory be corrected by incorporating radiative corrections and the q^2-dependence of the leading form factors, but there is no need yet to account for the g_3 form factors. We have seen in Chapter 7 that only when transition rates and $\alpha_{e\nu}$ coefficients are used can the Cabibbo parameters be fixed at reasonably steady values. If only rates are used those parameter are still poorly determined. Once the Cabibbo parameters are stable any further addition to the data becomes a real test of the Cabibbo theory. At the end of Chapter 7 we incorporated 7 new pieces of data that are not yet contained in the latest issue of the Review of Particle Properties. Two of them lead to strong discrepancies, 4 to 5 standard deviations each, with the Cabibbo theory; one because of the smallness of the error bars —the $\Sigma^- \rightarrow \Lambda e\nu$ rate—, and the other because of its central value —the α_e in $\Sigma^- \rightarrow ne\nu$.

The size of these discrepancies requires that the postulates of the Cabibbo theory be revised. This we have done in Chapter 8. The first natural thing to do is to incorporate small symmetry breaking in addition to the mass differences of the hyperons. This we did in a general way, because on the one hand theoretical calculations are not abundant nor reliable enough, and on the other hand because it would be useful to extract the symmetry breaking corrections from experiment. The result is that the discrepancy in the $\Sigma^- \rightarrow \Lambda e\nu$ rate can indeed be corrected with small symmetry breaking contributions, but the other discrepancy cannot, implying the presence of strong symmetry breaking —too strong for our first order expansions that correct the g_1 form factors to remain valid. We tried other possibilities also. Small admixtures of higher representations work just as well as small symmetry breaking, but cannot explain the second discrepancy either. Second class axial-vector currents help in no way and it is appealing to conclude that they are absent altogether.

Troublesome as it may seem, the discrepancy with α_e in $\Sigma^- \rightarrow ne\nu$ need not imply that the Cabibbo theory be abandoned. A proper definition of the symmetry limit from the outset may be all that is required. The Spectrum Generating approach, as we discussed in Chapter 9, provides such a possibility, by insisting that a complete set of commuting observables be defined from the beginning, in accordance with the basic postulates of Quantum Mechanics. This approach can be essentially reconciled with the Cabibbo theory. Although, small admixtures of higher representations in the axial-vector current are required. The Spectrum Generating approach opens the possibility to relate the Cabibbo angle to the mass differences between the hyperons. We gave an example of how this could be done with the multicommutators model. The current values of $\Xi^- \rightarrow \Sigma^0 e\nu$ and $\Xi^- \rightarrow \Lambda e\nu$ rates seem to rule out a simple version of this particular model. Nevertheless, the possibility of connecting the Cabibbo angle to mass differences remains attractive enough giving the Spectrum Generating approach a certain significance. As things stand today, only the Cabibbo-like Spectrum Generating model can explain satisfactorily the experimental evidence of hyperon semileptonic decays.

Before drawing long lasting conclusions one must await for better data. In particular the new values of the $\Sigma^- \to \Lambda e\nu$ rate and α_e in $\Sigma^- \to ne\nu$ should be confirmed. α_ν and α_e in $\Lambda \to pe\nu$ show some deviations. The one in α_ν could be very important, so a refined measurements of this quantity is also very necessary.

All of our comparisons between theory and experiment are limited by the low statistics of the currently available data. As new results appear it is clear that the comparision with theory must be updated. But the methods we have used certainly remain valid. In this respect one of our major aims has been to develop a framework that can be used in experiments with very high statistics —corresponding to several hundreds of thousands of events and even millions of events. The framework has been developed gradually over Chapters 2, 5 and 10, a little bit of Chapter 6, too. Some general results were given in the form of theorems and their consequences in Chapter 10. Contrary to what could be expected, it is not possible to uniquely determine all the form factors if no polarization information is available. Only F_1, G_1 and λ_{f_1} can be uniquely determined. The relative sign between F_1 and G_1 can be determined in the unpolarized Dalitz plot, but since they are combinations of the "little" form factors —the Dirac ones properly speaking— the relative sign of f_1 and g_1 is not well determined. For example, if g_1 is small then the presence of a not too small g_2 could reverse the sign of g_1. At any rate, whatever can be determined over the unpolarized Dalitz plot is probably going to be determined with very good precision. Then, some information from the polarized Dalitz plot can suffice to establish a unique solution. It is nevertheless necessary to have at our disposal good measurements of the polarized Dalitz plots, because with them several very refined tests of the V-A theory can be performed (which can look for a V+A admixture or the presence of small S, T, or P). Our framework is useful for model independent analysis of experiments that pretend to measure the relevant from factors in e-modes down to 1% and it can be extended in a straight forward —although tedious— fashion to higher precision. Its extension to μ-modes is also simple. For completeness, we give in Appendix 1 the exact numerical formulas for rates and asymmetries of all the HSD of Table 2.1. Also, for future reference, we give in Appendix 2 the predictions of the Cabibbo theory and Cabibbo-like Spectrum Generating model for all those decays not included in Chapters 8 and 9.

Important questions must be answered. To what extent is CVC reliable? Does the q^2-dependence of form factors behave as a pole or as a dipole?, to mention some of them. The answer to these detailed questions both theoretically and experimentally is necessary in helping to find the ultimate theory of strong and weak interactions, which should establish the extent to which SU(3) is broken in both the vector and axial-vector current parts. It is possible that some of the implications, as we mentioned above, be the incorporation of an Spectrum Generating approach in the Cabibbo theory. The answers will come with real high statistics experiments, with several hundreds of thousands of events. We only hope that the decade pattern of progress in β-decay is reduced in the near future to a year pattern.

Epilog

Let us close these lecture notes by elaborating the preface somewhat more. From the discovery of the Balmer series to the Schrodinger equation fifty years elapsed. In between, daring steps were taken by many people until Bohr formulated his theory. Quantum Mechanics textbooks tend to overemphasize the importance of the explanation of the Balmer series in Bohr theory. This was an important feature of this theory, but it was very much an ad-hoc explanation. The Bohr quantization of angular momentum was attrative because it was simple, but it was arbitrary and, in this sense, would not seem to go much farther than the phenomenological Balmer formula itself. One may wonder, then, what was it that prompted physicists of the time to take the Bohr theory so seriously. The answer, of course, is the prediction that the Bohr theory made for the Rydberg constant. What had been so far a phenomenological parameter, determined from experiment, was reproduced with very good accuracy by Bohr in terms of fundamental constants of nature. When something like this happens one is led to believe that there must be some truth in this kind of theories. As is well known later developments proved that the Bohr theory was not strictly correct.

It was Quantum Mechanics —with its uncertainty principle and angular momentum "whose square is $\ell(\ell+1)$ and not ℓ^2 ", and which in addition can be half-integer— that was finally established. Hardly can one believe that anyone forsaw the enormous changes implied by Quantum Mechanics. After the introduction of the Bohr theory only a decade or so was necessary to finally understand quantization. The reason why this was possible was the availability of fine structure measurements in Atomic Physics.

If a parallelism makes any sense at all, we may accept that in High Energy Physics we are close but not yet quite at the Bohr theory level. Today, in weak interactions we have several angles in addition to the Cabibbo one. They are all phenomenological and none has been related to fundamental constants of nature. It should be no surprise that changes similar to the passage from ℓ^2 to $\ell(\ell+1)$ may occur. But it also should not be a surprise that the equivalent of fine structure measurements are required.

From what we have seen in these lecture notes we may say that High Energy Physics is at the verge of its fine structure age. And it may well be that it will enter it through β decay When this occurs High Energy Physics will be at a new stage —quantitatively and qualitatively speaking. The last issue of Review of Particle Properties has printed in its back-cover the 1958 issue. The former is almost one inch thick. No doubt that High Energy Physics has come a long way in the last 25 years. But, no doubt either that it still has a long way to go.

Acknowledgments

The authors wish to express their gratitude to A. Bohm for encouraging them to write these lecture notes. They also want to thank him and E.C.G. Sudarshan for their hospitality on several ocassions at the Center for Particle Theory of the University of Texas at Austin. One of us wishes to acknowledge partial support of CONACyT (MEXICO) through project PCCBNA-020509.

References

Chapter 1.

1. For a review up to 1968 and older references see, for example, R.E. Marshak, Riazuddin, and C.P. Ryan, "Theory of Weak Interactions in Particle Physics" (Wiley-Interscience, New York, 1969).
2. For a review up to 1982 see, for example, M.A.B. Bég and A. Sirlin, "Gauge Theories of Weak Interactions. II", Physics Reports 88, 1(1982). A review at an intermediate level can be found in I.J. Aitchison and A.J.G. Hey, "Gauge Theories in Particle Physics" (Graduate Student Series in Physics, Adam Hilger LTD, Bristol, Ed. D.F. Brewer).

Chapter 2.

1. T.D. Lee and C.N. Yang, Phys. Rev 104, 254(1956).
2. E.C.G. Sudarshan and R.E. Marshak, Phys. Rev. 109, 1860(1958); R.P. Feyman and M. Gell-Mann, ibid. 109, 193(1958).
3. T.D. Lee and C.M. Yang 128, 855(1962).
4. S. Weinberg, Phys. Rev. Lett. 19, 1264(1967); and A. Salam, in Elementary Particle Theory: Relativistic Groups and Analyticity, Nobel Symposium No. 8 Edited by N. Svartholm (Admqvist and Wiksell, Stockholm, 1968).
5. J.D. Bjorken and S.D. Drell, "Relativistic Quantum Mechanics" (McGraw-Hill, New York, 1964).
6. S. Weinberg, Phys. Rev. 112, 1375(1958).
7. A. García, Phys. Rev. D25, 1348(1982), and Phys. Rev. D8, 1659(1983). In this text we have corrected some misprints that appear is the latter paper.
8. V. Linke, Nucl. Phys. B12, 669(1969).
9. D.R. Harrington, Phys. Rev. 120, 1482(1960).

Chapter 3.

1. W. Alles, Nuovo Cim. 26, 1429(1962).
2. A. García, Phys. Rev. D4, 2636(1971).
3. H. Pietschmann, Formulas and Results in Weak Interactions (Springer, Berlin, 1974).
4. S. Weinberg, Phys. Rev. 115, 481(1959).
5. I. Bender, V. Linke, and H.J. Rothe, Z. Physik 212, 190(1968).
6. A. García, Phys. Rev. D12, 2692(1975).
7. Review of Particle Properties, Particle Data Group, Phys. Lett 111B, April 1982.
8. P. Keller et al., Argonne-Chicago-Ohio State Collaboration, Phys. Rev. Lett. 48, 972(1982).
9. M. Bourquin et al. Z. Physik C12, 307(1982); WA2-Collaboration preprints,

"Measurement of hyperon semileptonic decays at the CERN super-proton synchrotron", CERN-EP/83-78, and CERN-EP/83-79, and IV Tests of the Cabibbo Model. The three preprints are submitted to Zeits. für Phys. C.

10. W. Tanenbaum et al. Phys. Rev. Lett. 33, 175(1974).
    W. Tanenbaum, Ph.D. Thesis, Yale University, 1974 (unpublished).
11. J. Lindquist et al. Phys. Rev. D16, 2104(1977).
12. For a review see M. Nagels et al., Nucl. Phys. B147, 189(1979).
13. Yu A. Mostovoi and A.I. Frank, Pis'ma Zh. Eksp. Teor. Fiz. 24, 43(1976). (English Translation: JETP Lett. 24, 38(1976)).

Chapter 4.

1. N. Cabibbo, Phys. Rev. Lett. 10, 531(1963); M. Gell-Mann and M. Lévy, Nuovo Cim. 16, 705(1960).
2. J.G. Kuryan, D. Lurie, and A.C. MacFarlane, J. Math. Phys. 6, 722(1965).
3. See, for example, H.M. Pilkuhn, Relativistic Physics, Texts and Monographs in Physics (Springer Verlag, Berlin, 1979).

Chapter 5.

1. A. Sirlin, Phys. Rev. 164, 1767(1967).
2. A. García and S.R. Juárez W., Phys. Rev. D22, 1132(1980), and references therein.
3. N. Meister and D.R. Yennie, Phys. Rev. 130, 1210(1963).
4. A. Sirlin, Rev. Mod. Phys. 50, 573(1978).
5. A. García, Phys. Lett. 105B, 224(1981).
6. D.A. Dicus and R.E. Norton, Phys. Rev. D1, 1360(1970).
7. A. García and A. Queijeiro, Phys. Rev. D27, 2101(1983).
8. A. Sirlin, Nucl. Physics B196, 83(1982).
9. See references in Ref. (1.9). See also R. Brown, in proceedings of the Brighton 1983 Conference of the European Physical Society.

Chapter 7.

1. A. García and P. Kielanowski, Phys. Rev. D26, 1090(1982). We have updated the analysis of this paper incorporating the new pieces of data.
2. R.E. Behrends and A. Sirlin, Phys. Rev. Lett. 4, 186(1960).
3. M. Ademollo and R. Gatto, Phys. Rev. Lett. 13, 264(1964).
4. J.F. Donoghue and B.R. Holstein, Phys. Rev. D25, 206(1982); and references therein.
5. J.F. Donoghue and B.R. Holstein, Phys. Rev. D25, 2015(1982).
6. A. García and P. Kielanowski, Phys. Lett. 110B, 498(1982).

Chapter 8.

1. A. García and P. Kielanowski, Phys. Rev. D26, 3100(1982). We have updated the analysis of this paper incorporating the new pieces of data.
2. S. Pakvasa, A. MacDonald and S.P. Rosen, Phys. Rev. 181, 1948(1969).
3. M. Kobayashi and T. Maskawa, Prog. Theor. Phys. 49, 652(1973).
4. R.E. Shrock and L.L. Wang, Phys. Rev. Lett. 41, 1692(1978).
   K. Kleinecht and B. Renk, Z. Physik C16, 7(1982).
5. A. Bohm, A. García and P. Kielanowski, Phys. Lett. 113B, 272(1982).

Chapter 9.

1. A. Bohm, Phys. Rev. 175, 1767(1968); A. Bohm and J. Werle, Nucl. Phys. B106, 165 (1976).
   A. Bohm, Phys. Rev. D13, 2110(1976), and references therein.
2. J. Werle, ICPT report, Trieste, 1965 (unpublished).
3. A. Bohm and R. Teese, J. Math. Phys. 18, 1434(1977).
4. P. Kielanowski, Lecture Notes in Physics 94, Group Theoretical Methods in Physics, proceedings of Austin Conference, 1978, edited by W. Beiglebock, A. Bohm, and E. Takasugi (Springer, Berlin 1979), p. 317; A. Bohm and P. Kielanowski, Phys. Rev. D27, 166(1983).
5. A. Bohm, P. Magnollay, A. García, and P. Kielanowski, Phys. Rev. D27, 180(1983).
6. A. Bohm and R.B. Teese, Phys. Rev. D26, 1103(1982).

Chapter 10.

1. A. García and P. Kielanowski, Phys. Lett. 120B, 214(1983); Phys. Lett. 121B, 424 (1983). In this text we have corrected some misprints and we have refined the analysis of the latter paper.
2. A. García and P. Kielanowski, submitted for publication.
3. Strictly speaking, the number of measurable functions or constants is eight, because D_3' and D_4' are each multiplied by the model independent radiative corrections $\phi + \theta$ and $\phi' + \theta'$. But, for e-modes and in view of the small difference between these two functions as shown in Eq. (5.34), it is only a common model independent function that practically matters, and then only seven functions or constants are measurable. This same observation applies to $\hat{D}_5'$ and $\hat{D}_6'$.

Appendix 1. Numerical Formulas for the Transition Rates and Angular Coefficients.

We have reserved this space to give complete numerical formulas for the transition rates and angular coefficients of the 16 e-mode and 10 μ-mode HSD. Our conventions are those of Chapter 2. The value of G_μ is given in Eq. (2.6). The different integrated observables are defined in Section 3.1, and the slope parameters of the leading form factors are defined in Eqs. (3.11) and (3.12). Radiative corrections are not included at all, nor are $\cos\theta$ or $\sin\theta$. The units are 10^{-3} sec^{-1} for the $n \to pe\nu$ rate and 10^{6} sec^{-1} for all others.

Tables 1A - 1D give the transition rate coefficients. Tables 2A - 2F give the coefficients of $R\alpha_{e\nu}$. Tables 3A - 3F give the coefficients of $R\alpha_e$. Tables 4A - 4F contain the coefficients of $R\alpha_\nu$. Tables 5A and 5B are for the coefficients of $R\alpha_{B'}$. Tables 6A - 6F give the coefficients of the asymmetry RA and in Tables 7A - 7F are the coefficients of RB. In the numerical integrations the complete phase space has been accounted for. The accuracy of the integrations is up to four significant digits.

The typing of the tables has been performed directly from the computer, in order to minimize copying errors. The integrations have been performed twice, starting from formulas (2.16) and (2.24) and from the results of Ref. (3.3) and (2.8); there was agreement all along, except for some misprints of Ref. (3.3), which we corrected by explicit evalution. The tables displayed here come from our formulas (2.16) and (2.24).

Table 1(A)

Process	f_1^2	f_1f_2	f_1f_3	f_2^2	f_3^2
$n\to pe\nu$	0.1897E+00	0.2761E-06	0.1350E-03	0.1839E-06	0.5612E-07
$\Sigma^+\to\Lambda e\nu$	0.2207E+00	0.7504E-03	0.3278E-05	0.4848E-03	0.4069E-07
$\Sigma^-\to\Lambda e\nu$	0.3654E+00	0.1510E-02	0.4863E-05	0.9725E-03	0.6644E-07
$\Lambda\to pe\nu$	0.1524E+02	0.3567E+00	0.9978E-04	0.2193E+00	0.3167E-05
$\Sigma^-\to ne\nu$	0.9039E+02	0.3986E+01	0.3766E-03	0.2379E+01	0.1616E-04
$\Xi^-\to\Lambda e\nu$	0.3219E+02	0.7220E+00	0.1535E-03	0.4446E+00	0.4771E-05
$\Xi^-\to\Sigma^0 e\nu$	0.3401E+01	0.2909E-01	0.2600E-04	0.1846E-01	0.5069E-06
$\Sigma^0\to pe\nu$	0.8456E+02	0.3652E+01	0.3589E-03	0.2182E+01	0.1525E-04
$\Xi^0\to\Sigma^+ e\nu$	0.2994E+01	0.2453E-01	0.2361E-04	0.1558E-01	0.4508E-06
$\Xi^-\to\Xi^0 e\nu$	0.1166E-05	0.2360E-10	0.1694E-09	0.1569E-10	0.1744E-12
$\Sigma^-\to\Sigma^0 e\nu$	0.2896E-06	0.4115E-11	0.5947E-10	0.2738E-11	0.5274E-13
$\Sigma^0\to\Sigma^+ e\nu$	0.2777E-07	0.1592E-12	0.8414E-11	0.1060E-12	0.5100E-14
$\Xi^-\to ne\nu$	0.5708E+03	0.4686E+02	0.1435E-02	0.2690E+02	0.8230E-04
$\Xi^0\to pe\nu$	0.5357E+03	0.4318E+02	0.1372E-02	0.2482E+02	0.7805E-04
$\Sigma^+\to ne\nu$	0.7789E+02	0.3261E+01	0.3375E-03	0.1952E+01	0.1413E-04
$\Xi^0\to\Sigma^- e\nu$	0.2176E+01	0.1559E-01	0.1833E-04	0.9932E-02	0.3278E-06
$\Lambda\to p\mu\nu$	0.2451E+01	0.4094E-01	0.3755E+00	0.2519E-01	0.2187E-01
$\Sigma^-\to n\mu\nu$	0.4000E+02	0.1570E+01	0.4657E+01	0.9380E+00	0.3071E+00
$\Xi^-\to\Lambda\mu\nu$	0.8707E+01	0.1565E+00	0.1046E+01	0.9645E-01	0.5534E-01
$\Xi^-\to\Sigma^0\mu\nu$	0.4403E-01	0.1411E-03	0.6553E-02	0.8957E-04	0.2813E-03
$\Sigma^0\to p\mu\nu$	0.3655E+02	0.1398E+01	0.4308E+01	0.8363E+00	0.2830E+00
$\Xi^0\to\Sigma^+\mu\nu$	0.2541E-01	0.6960E-04	0.3838E-02	0.4423E-04	0.1639E-03
$\Xi^-\to n\mu\nu$	0.3931E+03	0.3107E+02	0.3186E+02	0.1785E+02	0.2434E+01
$\Xi^0\to p\mu\nu$	0.3653E+03	0.2830E+02	0.3004E+02	0.1628E+02	0.2285E+01
$\Sigma^+\to n\mu\nu$	0.3265E+02	0.1204E+01	0.3900E+01	0.7211E+00	0.2543E+00
$\Xi^0\to\Sigma^-\mu\nu$	0.4130E-02	0.6568E-05	0.6394E-03	0.4187E-05	0.2666E-04

Table 1(B)

Process	$f_1\lambda_{f_1}$	$(f_1\lambda_{f_2}+\lambda_{f_1}f_2)$	$(f_1\lambda_{f_3}+\lambda_{f_1}f_3)$	$f_2\lambda_{f_2}$	$f_3\lambda_{f_3}$
$n\to pe\nu$	0.3697E-06	0.2915E-12	0.1545E-09	0.3885E-12	0.1420E-12
$\Sigma^+\to\Lambda e\nu$	0.4854E-03	0.1283E-05	0.5047E-08	0.1658E-05	0.1788E-09
$\Sigma^-\to\Lambda e\nu$	0.9737E-03	0.3128E-05	0.9072E-08	0.4029E-05	0.3539E-09
$\Lambda\to pe\nu$	0.2204E+00	0.4004E-02	0.1007E-05	0.4925E-02	0.9133E-07
$\Sigma^-\to ne\nu$	0.2402E+01	0.8205E-01	0.6957E-05	0.9801E-01	0.8541E-06
$\Xi^-\to\Lambda e\nu$	0.4467E+00	0.7776E-02	0.1486E-05	0.9579E-02	0.1320E-06
$\Xi^-\to\Sigma^0 e\nu$	0.1850E-01	0.1230E-03	0.9887E-07	0.1561E-03	0.5508E-08
$\Sigma^0\to pe\nu$	0.2203E+01	0.7370E-01	0.6500E-05	0.8813E-01	0.7901E-06
$\Xi^0\to\Sigma^+ e\nu$	0.1561E-01	0.9936E-04	0.8607E-07	0.1262E-03	0.4694E-08
$\Xi^-\to\Xi^0 e\nu$	0.1652E-10	0.2505E-15	0.1694E-14	0.3332E-15	0.4747E-17
$\Sigma^-\to\Sigma^0 e\nu$	0.2980E-11	0.3100E-16	0.4299E-15	0.4125E-16	0.1017E-17
$\Sigma^0\to\Sigma^+ e\nu$	0.1279E-12	0.5017E-18	0.2651E-16	0.6680E-18	0.4080E-19
$\Xi^-\to ne\nu$	0.2741E+02	0.1736E+01	0.4756E-04	0.1996E+01	0.7819E-05
$\Xi^0\to pe\nu$	0.2528E+02	0.1572E+01	0.4471E-04	0.1810E+01	0.7288E-05
$\Sigma^+\to ne\nu$	0.1970E+01	0.6389E-01	0.5934E-05	0.7652E-01	0.7107E-06
$\Xi^0\to\Sigma^- e\nu$	0.9948E-02	0.5537E-04	0.5861E-07	0.7057E-04	0.2994E-08
$\Lambda\to p\mu\nu$	0.8506E-01	0.7006E-03	0.6952E-02	0.8622E-03	0.8402E-03
$\Sigma^-\to n\mu\nu$	0.1952E+01	0.4110E-01	0.1322E+00	0.4912E-01	0.1916E-01
$\Xi^-\to\Lambda\mu\nu$	0.2600E+00	0.2369E-02	0.1723E-01	0.2919E-02	0.1931E-02
$\Xi^-\to\Sigma^0\mu\nu$	0.7289E-03	0.1134E-05	0.5486E-04	0.1439E-05	0.4739E-05
$\Sigma^0\to p\mu\nu$	0.1765E+01	0.3608E-01	0.1206E+00	0.4316E-01	0.1737E-01
$\Xi^0\to\Sigma^+\mu\nu$	0.4108E-03	0.5472E-06	0.3130E-04	0.6955E-06	0.2686E-05
$\Xi^-\to n\mu\nu$	0.2744E+02	0.1303E+01	0.1406E+01	0.1499E+01	0.2526E+00
$\Xi^0\to p\mu\nu$	0.2523E+02	0.1170E+01	0.1309E+01	0.1348E+01	0.2336E+00
$\Sigma^+\to n\mu\nu$	0.1549E+01	0.3036E-01	0.1068E+00	0.3638E-01	0.1522E-01
$\Xi^0\to\Sigma^-\mu\nu$	0.6136E-04	0.4781E-07	0.4766E-05	0.6096E-07	0.3981E-06

Table 1(C)

Process	g_1^2	g_1g_2	g_1g_3	g_2^2	g_3^2
$n \to pe\nu$	0.5692E+00	-0.1180E-02	-0.3681E-07	0.7202E-06	0.8853E-14
$\Sigma^+ \to \Lambda e\nu$	0.6613E+00	-0.5468E-01	-0.6292E-07	0.1453E-02	0.1787E-10
$\Sigma^- \to \Lambda e\nu$	0.1095E+01	-0.9962E-01	-0.1031E-06	0.2915E-02	0.3559E-10
$\Lambda \to pe\nu$	0.4533E+02	-0.9596E+01	-0.5186E-05	0.6535E+00	0.1016E-07
$\Sigma^- \to ne\nu$	0.2667E+03	-0.7639E+02	-0.2744E-04	0.7045E+01	0.1017E-06
$\Xi^- \to \Lambda e\nu$	0.9580E+02	-0.1987E+02	-0.7798E-05	0.1325E+01	0.1464E-07
$\Xi^- \to \Sigma^0 e\nu$	0.1017E+02	-0.1322E+01	-0.8003E-06	0.5526E-01	0.5718E-09
$\Sigma^0 \to pe\nu$	0.2496E+03	-0.7079E+02	-0.2586E-04	0.6464E+01	0.9383E-07
$\Xi^0 \to \Sigma^+ e\nu$	0.8957E+01	-0.1140E+01	-0.7108E-06	0.4663E-01	0.4862E-09
$\Xi^- \to \Xi^0 e\nu$	0.3497E-05	-0.2283E-07	-0.2378E-12	0.4760E-10	0.4378E-18
$\Sigma^- \to \Sigma^0 e\nu$	0.8687E-06	-0.4780E-08	-0.6859E-13	0.8372E-11	0.9232E-19
$\Sigma^0 \to \Sigma^+ e\nu$	0.8331E-07	-0.2972E-09	-0.5845E-14	0.3333E-12	0.3525E-20
$\Xi^- \to ne\nu$	0.1658E+04	-0.6359E+03	-0.1473E-03	0.7867E+02	0.1024E-05
$\Xi^0 \to pe\nu$	0.1557E+04	-0.5921E+03	-0.1394E-03	0.7262E+02	0.9514E-06
$\Sigma^+ \to ne\nu$	0.2300E+03	-0.6429E+02	-0.2391E-04	0.5784E+01	0.8408E-07
$\Xi^0 \to \Sigma^- e\nu$	0.6513E+01	-0.7762E+00	-0.5151E-06	0.2974E-01	0.3080E-09
$\Lambda \to p\mu\nu$	0.7318E+01	-0.1926E+01	-0.8704E-02	0.1405E+00	0.3927E-04
$\Sigma^- \to n\mu\nu$	0.1185E+03	-0.3862E+02	-0.2192E+00	0.3694E+01	0.1474E-02
$\Xi^- \to \Lambda\mu\nu$	0.2598E+02	-0.6436E+01	-0.2844E-01	0.4531E+00	0.1110E-03
$\Xi^- \to \Sigma^0 \mu\nu$	0.1320E+00	-0.2371E-01	-0.4028E-04	0.1112E-02	0.8470E-07
$\Sigma^0 \to p\mu\nu$	0.1083E+03	-0.3504E+02	-0.1986E+00	0.3321E+01	0.1318E-02
$\Xi^0 \to \Sigma^+ \mu\nu$	0.7617E-01	-0.1353E-01	-0.2030E-04	0.6240E-03	0.4178E-07
$\Xi^- \to n\mu\nu$	0.1146E+04	-0.4717E+03	-0.2565E+01	0.5941E+02	0.2664E-01
$\Xi^0 \to p\mu\nu$	0.1066E+04	-0.4355E+03	-0.2382E+01	0.5441E+02	0.2442E-01
$\Sigma^+ \to n\mu\nu$	0.9681E+02	-0.3097E+02	-0.1747E+00	0.2896E+01	0.1134E-02
$\Xi^0 \to \Sigma^- \mu\nu$	0.1239E-01	-0.2116E-02	-0.2020E-05	0.9249E-04	0.3843E-08

Table 1(D)

Process	$g_1\lambda_{g_1}$	$(g_1\lambda_{g_2}+\lambda_{g_1}g_2)$	$(g_1\lambda_{g_3}+\lambda_{g_1}g_3)$	$g_2\lambda_{g_2}$	$g_3\lambda_{g_3}$
$n \to pe\nu$	0.1330E-05	-0.1476E-08	-0.3539E-13	0.1838E-11	0.0000E+00
$\Sigma^+ \to \Lambda e\nu$	0.2422E-02	-0.1202E-03	-0.6925E-10	0.6624E-05	0.6109E-13
$\Sigma^- \to \Lambda e\nu$	0.4858E-02	-0.2653E-03	-0.1375E-09	0.1610E-04	0.1474E-12
$\Lambda \to pe\nu$	0.1089E+01	-0.1384E+00	-0.3742E-07	0.1955E-01	0.2282E-09
$\Sigma^- \to ne\nu$	0.1173E+02	-0.2019E+01	-0.3630E-06	0.3862E+00	0.4187E-08
$\Xi^- \to \Lambda e\nu$	0.2208E+01	-0.2749E+00	-0.5399E-07	0.3804E-01	0.3153E-09
$\Xi^- \to \Sigma^0 e\nu$	0.9209E-01	-0.7183E-02	-0.2176E-08	0.6227E-03	0.4834E-11
$\Sigma^0 \to pe\nu$	0.1077E+02	-0.1834E+01	-0.3353E-06	0.3474E+00	0.3787E-08
$\Xi^0 \to \Sigma^+ e\nu$	0.7772E-01	-0.5935E-02	-0.1852E-08	0.5037E-03	0.3939E-11
$\Xi^- \to \Xi^0 e\nu$	0.7999E-10	-0.3101E-12	-0.1747E-17	0.1340E-14	0.9324E-23
$\Sigma^- \to \Sigma^0 e\nu$	0.1414E-10	-0.4593E-13	-0.3685E-18	0.1666E-15	0.1398E-23
$\Sigma^0 \to \Sigma^+ e\nu$	0.5723E-12	-0.1182E-14	-0.1408E-19	0.2738E-17	0.2241E-25
$\Xi^- \to ne\nu$	0.1309E+03	-0.3021E+02	-0.3505E-05	0.7752E+01	0.7587E-07
$\Xi^0 \to pe\nu$	0.1209E+03	-0.2764E+02	-0.3261E-05	0.7034E+01	0.6928E-07
$\Sigma^+ \to ne\nu$	0.9634E+01	-0.1617E+01	-0.3010E-06	0.3018E+00	0.3294E-08
$\Xi^0 \to \Sigma^- e\nu$	0.4956E-01	-0.3544E-02	-0.1177E-08	0.2817E-03	0.2188E-11
$\Lambda \to p\mu\nu$	0.2721E+00	-0.3673E-01	-0.1446E-03	0.5410E-02	0.1355E-05
$\Sigma^- \to n\mu\nu$	0.6852E+01	-0.1191E+01	-0.5260E-02	0.2323E+00	0.7818E-04
$\Xi^- \to \Lambda\mu\nu$	0.8629E+00	-0.1113E+00	-0.4096E-03	0.1587E-01	0.3396E-05
$\Xi^- \to \Sigma^0 \mu\nu$	0.2211E-02	-0.1994E-03	-0.3223E-06	0.1873E-04	0.1363E-08
$\Sigma^0 \to p\mu\nu$	0.6172E+01	-0.1064E+01	-0.4710E-02	0.2054E+00	0.6891E-04
$\Xi^0 \to \Sigma^+ \mu\nu$	0.1243E-02	-0.1107E-03	-0.1591E-06	0.1023E-04	0.6578E-09
$\Xi^- \to n\mu\nu$	0.1057E+03	-0.2423E+02	-0.9115E-01	0.6270E+01	0.2265E-02
$\Xi^0 \to p\mu\nu$	0.9695E+02	-0.2204E+02	-0.8369E-01	0.5654E+01	0.2047E-02
$\Sigma^+ \to n\mu\nu$	0.5391E+01	-0.9169E+00	-0.4061E-02	0.1746E+00	0.5797E-04
$\Xi^0 \to \Sigma^- \mu\nu$	0.1847E-03	-0.1579E-04	-0.1469E-07	0.1381E-05	0.5597E-10

Table 2(A)

Process	f_1^2	$f_1 f_2$	$f_1 f_3$	f_2^2	$f_2 f_3$	f_3^2
$n \to pe\nu$	0.1384E+00	0.4315E-07	-0.2052E-06	0.2264E-08	0.1709E-10	-0.4111E-07
$\Sigma^+ \to \Lambda e\nu$	0.2063E+00	-0.3128E-03	-0.2059E-06	-0.2802E-03	0.5929E-12	-0.4320E-07
$\Sigma^- \to \Lambda e\nu$	0.3390E+00	-0.6425E-03	-0.3366E-06	-0.5700E-03	0.8707E-12	-0.7096E-07
$\Lambda \to pe\nu$	0.1249E+02	-0.1975E+00	-0.1632E-04	-0.1547E+00	0.2101E-10	-0.3667E-05
$\Sigma^- \to ne\nu$	0.6724E+02	-0.2532E+01	-0.8408E-04	-0.1859E+01	0.6939E-10	-0.1960E-04
$\Xi^- \to \Lambda e\nu$	0.2651E+02	-0.3963E+00	-0.2457E-04	-0.3117E+00	0.2306E-10	-0.5510E-05
$\Xi^- \to \Sigma^o e\nu$	0.3042E+01	-0.1357E-01	-0.2583E-05	-0.1153E-01	0.3865E-11	-0.5562E-06
$\Sigma^o \to pe\nu$	0.6315E+02	-0.2309E+01	-0.7931E-04	-0.1699E+01	0.6665E-10	-0.1847E-04
$\Xi^o \to \Sigma^+ e\nu$	0.2685E+01	-0.1137E-01	-0.2296E-05	-0.9685E-02	0.3542E-11	-0.4937E-06
$\Xi^- \to \Xi^o e\nu$	0.1132E-05	-0.6812E-11	-0.8519E-12	-0.6909E-11	0.1988E-16	-0.1711E-12
$\Sigma^- \to \Sigma^o e\nu$	0.2773E-06	-0.1062E-11	-0.2539E-12	-0.1102E-11	0.7995E-17	-0.5095E-13
$\Sigma^o \to \Sigma^+ e\nu$	0.2550E-07	-0.2773E-13	-0.2350E-13	-0.3219E-13	0.1000E-17	-0.4710E-14
$\Xi^- \to ne\nu$	0.3624E+03	-0.3487E+02	-0.4334E-03	-0.2373E+02	0.2189E-09	-0.1057E-03
$\Xi^o \to pe\nu$	0.3422E+03	-0.3197E+02	-0.4109E-03	-0.2181E+02	0.2113E-09	-0.1000E-03
$\Sigma^+ \to ne\nu$	0.5852E+02	-0.2046E+01	-0.7345E-04	-0.1511E+01	0.6298E-10	-0.1707E-04
$\Xi^o \to \Sigma^- e\nu$	0.1967E+01	-0.7093E-02	-0.1668E-05	-0.6096E-02	0.2746E-11	-0.3570E-06
$\Lambda \to p\mu\nu$	0.9437E+00	0.9594E-02	-0.6457E-01	-0.7913E-03	0.1070E-02	-0.1580E-01
$\Sigma^- \to n\mu\nu$	0.1832E+02	-0.1875E+00	-0.1173E+01	-0.2407E+00	0.1588E-01	-0.2925E+00
$\Xi^- \to \Lambda\mu\nu$	0.3943E+01	0.1822E-01	-0.1832E+00	-0.7797E-02	0.2446E-02	-0.4402E-01
$\Xi^- \to \Sigma^o \mu\nu$	0.1184E-01	0.1069E-03	-0.5086E-03	0.8193E-05	0.7779E-05	-0.1189E-03
$\Sigma^o \to p\mu\nu$	0.1667E+02	-0.1508E+00	-0.1073E+01	-0.2070E+00	0.1467E-01	-0.2675E+00
$\Xi^o \to \Sigma^+ \mu\nu$	0.6377E-02	0.5792E-04	-0.2761E-03	0.4061E-05	0.4271E-05	-0.6454E-04
$\Xi^- \to n\mu\nu$	0.1806E+03	-0.1334E+02	-0.1083E+02	-0.9496E+01	0.1116E+00	-0.2772E+01
$\Xi^o \to p\mu\nu$	0.1681E+03	-0.1189E+02	-0.1013E+02	-0.8513E+01	0.1056E+00	-0.2591E+01
$\Sigma^+ \to n\mu\nu$	0.1484E+02	-0.1114E+00	-0.9559E+00	-0.1698E+00	0.1319E-01	-0.2379E+00
$\Xi^o \to \Sigma^- \mu\nu$	0.8201E-03	0.7316E-05	-0.3526E-04	0.3592E-06	0.5537E-06	-0.8218E-05

Table 2(B)

Process	$f_1\lambda_{f_1}$	$(f_1\lambda_{f_2}+\lambda_{f_1}f_2)$	$(f_1\lambda_{f_3}+\lambda_{f_1}f_3)$	$f_2\lambda_{f_2}$	$(f_2\lambda_{f_3}+\lambda_{f_2}f_3)$	$f_3\lambda_{f_3}$
$n\rightarrow pe\nu$	0.8143E-07	-0.7792E-13	-0.7904E-10	-0.1521E-12	0.1671E-16	-0.1565E-12
$\Sigma^+\rightarrow\Lambda e\nu$	-0.4647E-04	-0.1382E-05	-0.5357E-08	-0.1980E-05	0.4549E-15	-0.2764E-09
$\Sigma^-\rightarrow\Lambda e\nu$	-0.1030E-03	-0.3394E-05	-0.9687E-08	-0.4841E-05	0.8049E-15	-0.5486E-09
$\Lambda\rightarrow pe\nu$	-0.5674E-01	-0.4799E-02	-0.1166E-05	-0.6416E-02	0.9895E-13	-0.1474E-06
$\Sigma^-\rightarrow ne\nu$	-0.8593E+00	-0.1040E+00	-0.8441E-05	-0.1338E+00	0.5777E-12	-0.1411E-05
$\Xi^-\rightarrow\Lambda e\nu$	-0.1125E+00	-0.9289E-02	-0.1716E-05	-0.1245E-01	0.1043E-12	-0.2129E-06
$\Xi^-\rightarrow\Sigma^0 e\nu$	-0.2837E-02	-0.1380E-03	-0.1085E-06	-0.1927E-03	0.7119E-14	-0.8655E-08
$\Sigma^0\rightarrow pe\nu$	-0.7795E+00	-0.9326E-01	-0.7873E-05	-0.1201E+00	0.5447E-12	-0.1304E-05
$\Xi^0\rightarrow\Sigma^+ e\nu$	-0.2342E-02	-0.1112E-03	-0.9426E-07	-0.1555E-03	0.6263E-14	-0.7369E-08
$\Xi^-\rightarrow\Xi^0 e\nu$	0.2220E-12	-0.2405E-15	-0.1623E-14	-0.3594E-15	0.1235E-21	-0.7035E-17
$\Sigma^-\rightarrow\Sigma^0 e\nu$	0.8377E-13	-0.2875E-16	-0.3994E-15	-0.4299E-16	0.3762E-22	-0.1491E-17
$\Sigma^0\rightarrow\Sigma^+ e\nu$	0.8924E-14	-0.4128E-18	-0.2249E-16	-0.6204E-18	0.2241E-23	-0.5772E-19
$\Xi^-\rightarrow ne\nu$	-0.1357E+02	-0.2356E+01	-0.6105E-04	-0.2882E+01	0.3119E-11	-0.1328E-04
$\Xi^0\rightarrow pe\nu$	-0.1240E+02	-0.2129E+01	-0.5730E-04	-0.2609E+01	0.2964E-11	-0.1237E-04
$\Sigma^+\rightarrow ne\nu$	-0.6858E+00	-0.8060E-01	-0.7169E-05	-0.1040E+00	0.5007E-12	-0.1172E-05
$\Xi^0\rightarrow\Sigma^- e\nu$	-0.1393E-02	-0.6156E-04	-0.6383E-07	-0.8647E-04	0.4276E-14	-0.4687E-08
$\Lambda\rightarrow p\mu\nu$	0.8897E-02	0.1379E-04	-0.3053E-02	-0.1989E-03	0.1794E-04	-0.7823E-03
$\Sigma^-\rightarrow n\mu\nu$	-0.1200E+00	-0.2196E-01	-0.9761E-01	-0.3179E-01	0.3674E-03	-0.2457E-01
$\Xi^-\rightarrow\Lambda\mu\nu$	0.2023E-01	-0.4311E-03	-0.9196E-02	-0.1049E-02	0.3525E-04	-0.2053E-02
$\Xi^-\rightarrow\Sigma^0\mu\nu$	0.1374E-03	0.8081E-06	-0.8395E-05	0.6318E-07	0.6318E-07	-0.2310E-05
$\Sigma^0\rightarrow p\mu\nu$	-0.9647E-01	-0.1863E-01	-0.8783E-01	-0.2727E-01	0.3361E-03	-0.2210E-01
$\Xi^0\rightarrow\Sigma^+\mu\nu$	0.7584E-04	0.4368E-06	-0.4189E-05	0.3718E-07	0.3394E-07	-0.1201E-05
$\Xi^-\rightarrow n\mu\nu$	-0.7729E+01	-0.1268E+01	-0.1411E+01	-0.1572E+01	0.3510E-02	-0.3845E+00
$\Xi^0\rightarrow p\mu\nu$	-0.6915E+01	-0.1124E+01	-0.1303E+01	-0.1397E+01	0.3294E-02	-0.3541E+00
$\Sigma^+\rightarrow n\mu\nu$	-0.7044E-01	-0.1496E-01	-0.7632E-01	-0.2226E-01	0.2972E-03	-0.1915E-01
$\Xi^0\rightarrow\Sigma^-\mu\nu$	0.1011E-04	0.5270E-07	-0.4116E-06	0.4105E-08	0.4065E-08	-0.1336E-06

Table 2(C)

Process	g_1^2	g_1g_2	g_1g_3	g_2^2	g_2g_3	g_3^2
$n \to pe\nu$	-0.1395E+00	0.7656E-03	-0.4100E-07	-0.5679E-06	0.1709E-10	0.7958E-16
$\Sigma^+ \to \Lambda e\nu$	-0.2617E+00	0.5805E-01	-0.3799E-07	-0.1777E-02	0.5929E-12	-0.7449E-11
$\Sigma^- \to \Lambda e\nu$	-0.4401E+00	0.1064E+00	-0.6156E-07	-0.3581E-02	0.8707E-12	-0.1515E-10
$\Lambda \to pe\nu$	-0.2242E+02	0.1111E+02	-0.2569E-05	-0.8561E+00	0.2101E-10	-0.5626E-08
$\Sigma^- \to ne\nu$	-0.1473E+03	0.9268E+02	-0.1175E-04	-0.9576E+01	0.6939E-10	-0.6460E-07
$\Xi^- \to \Lambda e\nu$	-0.4706E+02	0.2294E+02	-0.3892E-05	-0.1732E+01	0.2306E-10	-0.8035E-08
$\Xi^- \to \Sigma^o e\nu$	-0.4392E+01	0.1451E+01	-0.4521E-06	-0.6936E-01	0.3865E-11	-0.2667E-09
$\Sigma^o \to pe\nu$	-0.1373E+03	0.8574E+02	-0.1114E-04	-0.8775E+01	0.6665E-10	-0.5931E-07
$\Xi^o \to \Sigma^+ e\nu$	-0.3849E+01	0.1248E+01	-0.4031E-06	-0.5845E-01	0.3542E-11	-0.2254E-09
$\Xi^- \to \Xi^o e\nu$	-0.1155E-05	0.2223E-07	-0.1694E-12	-0.5411E-10	0.1988E-16	-0.1308E-18
$\Sigma^- \to \Sigma^o e\nu$	-0.2821E-06	0.4561E-08	-0.5054E-13	-0.9337E-11	0.7995E-17	-0.2538E-19
$\Sigma^o \to \Sigma^+ e\nu$	-0.2580E-07	0.2667E-09	-0.4686E-14	-0.3512E-12	0.1000E-17	-0.7451E-21
$\Xi^- \to ne\nu$	-0.1040E+04	0.8164E+03	-0.4943E-04	-0.1118E+03	0.2189E-09	-0.7619E-06
$\Xi^o \to pe\nu$	-0.9731E+03	0.7588E+03	-0.4724E-04	-0.1031E+03	0.2113E-09	-0.7044E-06
$\Sigma^+ \to ne\nu$	-0.1258E+03	0.7767E+02	-0.1039E-04	-0.7836E+01	0.6298E-10	-0.5276E-07
$\Xi^o \to \Sigma^- e\nu$	-0.2759E+01	0.8454E+00	-0.2955E-06	-0.3711E-01	0.2746E-11	-0.1402E-09
$\Lambda \to p\mu\nu$	-0.2579E+01	0.1184E+01	-0.1159E-01	-0.1038E+00	0.1070E-02	-0.2429E-05
$\Sigma^- \to n\mu\nu$	-0.5671E+02	0.3352E+02	-0.1858E+00	-0.3682E+01	0.1588E-01	-0.4073E-03
$\Xi^- \to \Lambda\mu\nu$	-0.9822E+01	0.4471E+01	-0.3226E-01	-0.3726E+00	0.2446E-02	-0.1217E-04
$\Xi^- \to \Sigma^o \mu\nu$	-0.2535E-01	0.7763E-02	-0.9895E-04	-0.4720E-03	0.7779E-05	0.6648E-08
$\Sigma^o \to p\mu\nu$	-0.5138E+02	0.3012E+02	-0.1710E+00	-0.3283E+01	0.1467E-01	-0.3534E-03
$\Xi^o \to \Sigma^+ \mu\nu$	-0.1361E-01	0.4093E-02	-0.5393E-04	-0.2465E-03	0.4271E-05	0.3378E-08
$\Xi^- \to n\mu\nu$	-0.6821E+03	0.5118E+03	-0.1396E+01	-0.7202E+02	0.1116E+00	-0.1377E-01
$\Xi^o \to p\mu\nu$	-0.6305E+03	0.4698E+03	-0.1316E+01	-0.6561E+02	0.1056E+00	-0.1244E-01
$\Sigma^+ \to n\mu\nu$	-0.4537E+02	0.2628E+02	-0.1533E+00	-0.2830E+01	0.1319E-01	-0.2918E-03
$\Xi^o \to \Sigma^- \mu\nu$	-0.1725E-02	0.4904E-03	-0.6953E-05	-0.2855E-04	0.5537E-06	0.3070E-09

Table 2(D)

Process	$g_1\lambda_{g_1}$	$(g_1\lambda_{g_2}+\lambda_{g_1}g_2)$	$(g_1\lambda_{g_3}+\lambda_{g_1}g_3)$	$g_2\lambda_{g_2}$	$(g_2\lambda_{g_3}+\lambda_{g_2}g_3)$	$g_3\lambda_{g_3}$
$n\to pe\nu$	-0.9766E-06	0.1536E-08	-0.2384E-13	-0.2110E-11	0.1671E-16	0.0000E+00
$\Sigma^+\to\Lambda e\nu$	-0.3041E-02	0.1857E-03	0.2884E-10	-0.1082E-04	0.4549E-15	-0.6581E-13
$\Sigma^-\to\Lambda e\nu$	-0.6125E-02	0.4112E-03	0.5846E-10	-0.2636E-04	0.8049E-15	-0.1600E-12
$\Lambda\to pe\nu$	-0.1460E+01	0.2234E+00	0.2071E-07	-0.3309E-01	0.9895E-13	-0.2735E-09
$\Sigma^-\to ne\nu$	-0.1629E+02	0.3335E+01	0.2306E-06	-0.6661E+00	0.5777E-12	-0.5309E-08
$\Xi^-\to\Lambda e\nu$	-0.2954E+01	0.4432E+00	0.2963E-07	-0.6431E-01	0.1043E-12	-0.3766E-09
$\Xi^-\to\Sigma^o e\nu$	-0.1185E+00	0.1129E-01	0.1015E-08	-0.1031E-02	0.7119E-14	-0.5423E-11
$\Sigma^o\to pe\nu$	-0.1493E+02	0.3028E+01	0.2120E-06	-0.5988E+00	0.5447E-12	-0.4792E-08
$\Xi^o\to\Sigma^+ e\nu$	-0.9987E-01	0.9317E-02	0.8583E-09	-0.8333E-03	0.6263E-14	-0.4410E-11
$\Xi^-\to\Xi^o e\nu$	-0.9385E-10	0.4587E-12	0.4255E-18	-0.2108E-14	0.1235E-21	-0.9000E-23
$\Sigma^-\to\Sigma^o e\nu$	-0.1628E-10	0.6711E-13	0.6873E-19	-0.2591E-15	0.3762E-22	-0.1309E-23
$\Sigma^o\to\Sigma^+ e\nu$	-0.6197E-12	0.1657E-14	0.3766E-21	-0.4103E-17	0.2241E-23	-0.1896E-25
$\Xi^-\to ne\nu$	-0.1898E+03	0.5131E+02	0.2607E-05	-0.1367E+02	0.3119E-11	-0.1030E-06
$\Xi^o\to pe\nu$	-0.1750E+03	0.4692E+02	0.2414E-05	-0.1240E+02	0.2964E-11	-0.9382E-07
$\Sigma^+\to ne\nu$	-0.1333E+02	0.2666E+01	0.1889E-06	-0.5196E+00	0.5007E-12	-0.4156E-08
$\Xi^o\to\Sigma^- e\nu$	-0.6342E-01	0.5549E-02	0.5354E-09	-0.4649E-03	0.4276E-14	-0.2433E-11
$\Lambda\to p\mu\nu$	-0.1654E+00	0.3078E-01	-0.1611E-03	-0.5116E-02	0.1794E-04	-0.3590E-06
$\Sigma^-\to n\mu\nu$	-0.6418E+01	0.1456E+01	-0.1965E-02	-0.3061E+00	0.3674E-03	-0.5219E-04
$\Xi^-\to\Lambda\mu\nu$	-0.6214E+00	0.1092E+00	-0.3360E-03	-0.1725E-01	0.3525E-04	-0.1330E-05
$\Xi^-\to\Sigma^o\mu\nu$	-0.5852E-03	0.7886E-04	-0.7840E-06	-0.9165E-05	0.6318E-07	0.4058E-10
$\Sigma^o\to p\mu\nu$	-0.5714E+01	0.1288E+01	-0.1864E-02	-0.2685E+00	0.3361E-03	-0.4502E-04
$\Xi^o\to\Sigma^+\mu\nu$	-0.2962E-03	0.3971E-04	-0.4206E-06	-0.4586E-05	0.3394E-07	0.2738E-10
$\Xi^-\to n\mu\nu$	-0.1273E+03	0.3616E+02	0.1488E-01	-0.9848E+01	0.3510E-02	-0.2341E-02
$\Xi^o\to p\mu\nu$	-0.1159E+03	0.3272E+02	0.1237E-01	-0.8842E+01	0.3294E-02	-0.2095E-02
$\Sigma^+\to n\mu\nu$	-0.4915E+01	0.1096E+01	-0.1725E-02	-0.2256E+00	0.2972E-03	-0.3679E-04
$\Xi^o\to\Sigma^-\mu\nu$	-0.3128E-04	0.4112E-05	-0.5068E-07	-0.4642E-06	0.4065E-08	0.3421E-11

Table 2(E)

Process	f_1g_1	f_1g_2	f_2g_1	f_2g_2
$n \to pe\nu$	-0.1155E-03	0.1591E-06	-0.2309E-03	0.3179E-06
$\Sigma^+ \to \Lambda e\nu$	-0.6423E-05	0.3984E-06	-0.1245E-04	0.7720E-06
$\Sigma^- \to \Lambda e\nu$	-0.9561E-05	0.6527E-06	-0.1847E-04	0.1261E-05
$\Lambda \to pe\nu$	-0.2002E-03	0.3183E-04	-0.3687E-03	0.5860E-04
$\Sigma^- \to ne\nu$	-0.7620E-03	0.1640E-03	-0.1360E-02	0.2928E-03
$\Xi^- \to \Lambda e\nu$	-0.3084E-03	0.4801E-04	-0.5688E-03	0.8855E-04
$\Xi^- \to \Sigma^0 e\nu$	-0.5168E-04	0.5040E-05	-0.9832E-04	0.9589E-05
$\Sigma^0 \to pe\nu$	-0.7259E-03	0.1547E-03	-0.1297E-02	0.2765E-03
$\Xi^0 \to \Sigma^+ e\nu$	-0.4691E-04	0.4479E-05	-0.8934E-04	0.8530E-05
$\Xi^- \to \Xi^0 e\nu$	-0.2659E-09	0.1292E-11	-0.5304E-09	0.2577E-11
$\Sigma^- \to \Sigma^0 e\nu$	-0.8778E-10	0.3578E-12	-0.1752E-09	0.7141E-12
$\Sigma^0 \to \Sigma^+ e\nu$	-0.1089E-10	0.2831E-13	-0.2175E-10	0.5655E-13
$\Xi^- \to ne\nu$	-0.2927E-02	0.8457E-03	-0.5009E-02	0.1447E-02
$\Xi^0 \to pe\nu$	-0.2799E-02	0.8016E-03	-0.4796E-02	0.1374E-02
$\Sigma^+ \to ne\nu$	-0.6823E-03	0.1433E-03	-0.1221E-02	0.2565E-03
$\Xi^0 \to \Sigma^- e\nu$	-0.3637E-04	0.3251E-05	-0.6948E-04	0.6212E-05
$\Lambda \to p\mu\nu$	-0.2386E+00	0.3792E-01	-0.4393E+00	0.6982E-01
$\Sigma^- \to n\mu\nu$	-0.4079E+01	0.8781E+00	-0.7279E+01	0.1567E+01
$\Xi^- \to \Lambda\mu\nu$	-0.7652E+00	0.1191E+00	-0.1411E+01	0.2197E+00
$\Xi^- \to \Sigma^0\mu\nu$	-0.2433E-02	0.2373E-03	-0.4629E-02	0.4514E-03
$\Sigma^0 \to p\mu\nu$	-0.3737E+01	0.7966E+00	-0.6678E+01	0.1423E+01
$\Xi^0 \to \Sigma^+\mu\nu$	-0.1323E-02	0.1263E-03	-0.2519E-02	0.2405E-03
$\Xi^- \to n\mu\nu$	-0.3491E+02	0.1009E+02	-0.5974E+02	0.1726E+02
$\Xi^0 \to p\mu\nu$	-0.3271E+02	0.9368E+01	-0.5605E+02	0.1605E+02
$\Sigma^+ \to n\mu\nu$	-0.3342E+01	0.7019E+00	-0.5982E+01	0.1256E+01
$\Xi^0 \to \Sigma^-\mu\nu$	-0.1715E-03	0.1533E-04	-0.3277E-03	0.2930E-04

Table 2(F)

Process	$(f_1\lambda_{g_1}+\lambda_{f_1}g_1)$	$(f_1\lambda_{g_2}+\lambda_{f_1}g_2)$	$(f_2\lambda_{g_1}+\lambda_{f_2}g_1)$	$(f_2\lambda_{g_2}+\lambda_{f_2}g_2)$
$n \to pe\nu$	-0.1130E-09	0.1555E-12	-0.2258E-09	0.3108E-12
$\Sigma^+ \to \Lambda e\nu$	-0.4929E-08	0.3057E-09	-0.9552E-08	0.5924E-09
$\Sigma^- \to \Lambda e\nu$	-0.8839E-08	0.6034E-09	-0.1707E-07	0.1166E-08
$\Lambda \to pe\nu$	-0.9432E-06	0.1499E-06	-0.1736E-05	0.2760E-06
$\Sigma^- \to ne\nu$	-0.6343E-05	0.1366E-05	-0.1132E-04	0.2437E-05
$\Xi^- \to \Lambda e\nu$	-0.1394E-05	0.2171E-06	-0.2571E-05	0.4003E-06
$\Xi^- \to \Sigma^0 e\nu$	-0.9519E-07	0.9283E-08	-0.1811E-06	0.1766E-07
$\Sigma^0 \to pe\nu$	-0.5933E-05	0.1265E-05	-0.1060E-04	0.2260E-05
$\Xi^0 \to \Sigma^+ e\nu$	-0.8293E-07	0.7918E-08	-0.1579E-06	0.1508E-07
$\Xi^- \to \Xi^0 e\nu$	-0.1651E-14	0.8023E-17	-0.3295E-14	0.1601E-16
$\Sigma^- \to \Sigma^0 e\nu$	-0.4130E-15	0.1683E-17	-0.8244E-15	0.3360E-17
$\Sigma^0 \to \Sigma^+ e\nu$	-0.2440E-16	0.6344E-19	-0.4874E-16	0.1267E-18
$\Xi^- \to ne\nu$	-0.4171E-04	0.1205E-04	-0.7137E-04	0.2062E-04
$\Xi^0 \to pe\nu$	-0.3926E-04	0.1124E-04	-0.6727E-04	0.1927E-04
$\Sigma^+ \to ne\nu$	-0.5425E-05	0.1139E-05	-0.9711E-05	0.2039E-05
$\Xi^0 \to \Sigma^- e\nu$	-0.5662E-07	0.5062E-08	-0.1082E-06	0.9672E-08
$\Lambda \to p\mu\nu$	-0.3999E-02	0.6356E-03	-0.7363E-02	0.1170E-02
$\Sigma^- \to n\mu\nu$	-0.9435E-01	0.2031E-01	-0.1684E+00	0.3625E-01
$\Xi^- \to \Lambda\mu\nu$	-0.1102E-01	0.1716E-02	-0.2033E-01	0.3166E-02
$\Xi^- \to \Sigma^0\mu\nu$	-0.1976E-04	0.1927E-05	-0.3759E-04	0.3666E-05
$\Sigma^0 \to p\mu\nu$	-0.8563E-01	0.1825E-01	-0.1530E+00	0.3261E-01
$\Xi^0 \to \Sigma^+\mu\nu$	-0.1051E-04	0.1004E-05	-0.2002E-04	0.1911E-05
$\Xi^- \to n\mu\nu$	-0.1098E+01	0.3172E+00	-0.1879E+01	0.5427E+00
$\Xi^0 \to p\mu\nu$	-0.1020E+01	0.2922E+00	-0.1748E+01	0.5007E+00
$\Sigma^+ \to n\mu\nu$	-0.7531E-01	0.1582E-01	-0.1348E+00	0.2831E-01
$\Xi^0 \to \Sigma^-\mu\nu$	-0.1259E-05	0.1126E-06	-0.2405E-05	0.2151E-06

Table 3(A)

Process	f_1^2	$f_1 f_2$	f_2^2
$n \to pe\nu$	-0.4446E-04	-0.8887E-04	-0.2401E-07
$\Sigma^+ \to \Lambda e\nu$	-0.4810E-02	-0.9411E-02	-0.1718E-03
$\Sigma^- \to \Lambda e\nu$	-0.8814E-02	-0.1721E-01	-0.3469E-03
$\Lambda \to pe\nu$	-0.9262E+00	-0.1752E+01	-0.8568E-01
$\Sigma^- \to ne\nu$	-0.7814E+01	-0.1450E+02	-0.9833E+00
$\Xi^- \to \Lambda e\nu$	-0.1911E+01	-0.3619E+01	-0.1731E+00
$\Xi^- \to \Sigma^o e\nu$	-0.1202E+00	-0.2323E+00	-0.6783E-02
$\Sigma^o \to pe\nu$	-0.7225E+01	-0.1341E+02	-0.9001E+00
$\Xi^o \to \Sigma^+ e\nu$	-0.1034E+00	-0.2000E+00	-0.5712E-02
$\Xi^- \to \Xi^o e\nu$	-0.1806E-08	-0.3606E-08	-0.4884E-11
$\Sigma^- \to \Sigma^o e\nu$	-0.3651E-09	-0.7291E-09	-0.8145E-12
$\Sigma^o \to \Sigma^+ e\nu$	-0.2039E-10	-0.4075E-10	-0.2762E-13
$\Xi^- \to ne\nu$	-0.7059E+02	-0.1278E+03	-0.1197E+02
$\Xi^o \to pe\nu$	-0.6554E+02	-0.1187E+03	-0.1101E+02
$\Sigma^+ \to ne\nu$	-0.6540E+01	-0.1215E+02	-0.8025E+00
$\Xi^o \to \Sigma^- e\nu$	-0.7004E-01	-0.1357E+00	-0.3619E-02
$\Lambda \to p\mu\nu$	-0.4027E-01	-0.7547E-01	-0.2433E-02
$\Sigma^- \to n\mu\nu$	-0.1725E+01	-0.3181E+01	-0.1821E+00
$\Xi^- \to \Lambda\mu\nu$	-0.1877E+00	-0.3528E+00	-0.1217E-01
$\Xi^- \to \Sigma^o \mu\nu$	-0.1174E-03	-0.2246E-03	-0.2386E-05
$\Sigma^o \to p\mu\nu$	-0.1533E+01	-0.2828E+01	-0.1592E+00
$\Xi^o \to \Sigma^+ \mu\nu$	-0.5431E-04	-0.1040E-03	-0.1004E-05
$\Xi^- \to n\mu\nu$	-0.3362E+02	-0.6066E+02	-0.5372E+01
$\Xi^o \to p\mu\nu$	-0.3065E+02	-0.5534E+02	-0.4843E+01
$\Sigma^+ \to n\mu\nu$	-0.1318E+01	-0.2435E+01	-0.1338E+00
$\Xi^o \to \Sigma^- \mu\nu$	-0.4141E-05	-0.7941E-05	-0.5663E-07

Table 3(B)

Process	$f_1\lambda_{f_1}$	$(f_1\lambda_{f_2}+\lambda_{f_1}f_2)$	$f_2\lambda_{f_2}$
$n \to pe\nu$	-0.1025E-09	-0.1025E-09	-0.6534E-13
$\Sigma^+ \to \Lambda e\nu$	-0.1584E-04	-0.1552E-04	-0.6611E-06
$\Sigma^- \to \Lambda e\nu$	-0.3516E-04	-0.3438E-04	-0.1617E-05
$\Lambda \to pe\nu$	-0.1997E-01	-0.1897E-01	-0.2163E-02
$\Sigma^- \to ne\nu$	-0.3086E+00	-0.2881E+00	-0.4552E-01
$\Xi^- \to \Lambda e\nu$	-0.3955E-01	-0.3761E-01	-0.4195E-02
$\Xi^- \to \Sigma^o e\nu$	-0.9778E-03	-0.9471E-03	-0.6451E-04
$\Sigma^o \to pe\nu$	-0.2797E+00	-0.2613E+00	-0.4085E-01
$\Xi^o \to \Sigma^+ e\nu$	-0.8064E-03	-0.7816E-03	-0.5207E-04
$\Xi^- \to \Xi^o e\nu$	-0.3787E-13	-0.3781E-13	-0.1206E-15
$\Sigma^- \to \Sigma^o e\nu$	-0.5497E-14	-0.5489E-14	-0.1449E-16
$\Sigma^o \to \Sigma^+ e\nu$	-0.1322E-15	-0.1321E-15	-0.2130E-18
$\Xi^- \to ne\nu$	-0.5007E+01	-0.4575E+01	-0.9967E+00
$\Xi^o \to pe\nu$	-0.4570E+01	-0.4178E+01	-0.9016E+00
$\Sigma^+ \to ne\nu$	-0.2458E+00	-0.2299E+00	-0.3536E-01
$\Xi^o \to \Sigma^- e\nu$	-0.4789E-03	-0.4651E-03	-0.2893E-04
$\Lambda \to p\mu\nu$	-0.1492E-02	-0.1401E-02	-0.9895E-04
$\Sigma^- \to n\mu\nu$	-0.9832E-01	-0.9104E-01	-0.1179E-01
$\Xi^- \to \Lambda\mu\nu$	-0.6193E-02	-0.5833E-02	-0.4500E-03
$\Xi^- \to \Sigma^o \mu\nu$	-0.1966E-05	-0.1881E-05	-0.4156E-07
$\Sigma^o \to p\mu\nu$	-0.8615E-01	-0.7981E-01	-0.1016E-01
$\Xi^o \to \Sigma^+ \mu\nu$	-0.8859E-06	-0.8481E-06	-0.1692E-07
$\Xi^- \to n\mu\nu$	-0.2999E+01	-0.2728E+01	-0.5528E+00
$\Xi^o \to p\mu\nu$	-0.2699E+01	-0.2456E+01	-0.4917E+00
$\Sigma^+ \to n\mu\nu$	-0.7246E-01	-0.6718E-01	-0.8339E-02
$\Xi^o \to \Sigma^- \mu\nu$	-0.6174E-07	-0.5920E-07	-0.8594E-09

Table 3(C)

Process	g_1^2	g_1g_2	g_2^2
$n\to pe\nu$	-0.2775E+00	0.6753E-03	-0.4037E-06
$\Sigma^+\to\Lambda e\nu$	-0.4454E+00	0.4600E-01	-0.1140E-02
$\Sigma^-\to\Lambda e\nu$	-0.7379E+00	0.8390E-01	-0.2289E-02
$\Lambda\to pe\nu$	-0.3101E+02	0.8200E+01	-0.5199E+00
$\Sigma^-\to ne\nu$	-0.1841E+03	0.6588E+02	-0.5649E+01
$\Xi^-\to\Lambda e\nu$	-0.6551E+02	0.1697E+02	-0.1054E+01
$\Xi^-\to\Sigma^0 e\nu$	-0.6891E+01	0.1119E+01	-0.4358E-01
$\Sigma^0\to pe\nu$	-0.1723E+03	0.6103E+02	-0.5182E+01
$\Xi^0\to\Sigma^+ e\nu$	-0.6066E+01	0.9642E+00	-0.3677E-01
$\Xi^-\to\Xi^0 e\nu$	-0.2278E-05	0.1853E-07	-0.3628E-10
$\Sigma^-\to\Sigma^0 e\nu$	-0.5575E-06	0.3816E-08	-0.6292E-11
$\Sigma^0\to\Sigma^+ e\nu$	-0.5117E-07	0.2254E-09	-0.2400E-12
$\Xi^-\to ne\nu$	-0.1157E+04	0.5550E+03	-0.6373E+02
$\Xi^0\to pe\nu$	-0.1087E+04	0.5165E+03	-0.5881E+02
$\Sigma^+\to ne\nu$	-0.1587E+03	0.5539E+02	-0.4635E+01
$\Xi^0\to\Sigma^- e\nu$	-0.4406E+01	0.6559E+00	-0.2343E-01
$\Lambda\to p\mu\nu$	-0.2877E+01	0.8416E+00	-0.6110E-01
$\Sigma^-\to n\mu\nu$	-0.6020E+02	0.2294E+02	-0.2149E+01
$\Xi^-\to\Lambda\mu\nu$	-0.1145E+02	0.3225E+01	-0.2246E+00
$\Xi^-\to\Sigma^0\mu\nu$	-0.3180E-01	0.5981E-02	-0.2808E-03
$\Sigma^0\to p\mu\nu$	-0.5464E+02	0.2064E+02	-0.1918E+01
$\Xi^0\to\Sigma^+\mu\nu$	-0.1710E-01	0.3162E-02	-0.1460E-03
$\Xi^-\to n\mu\nu$	-0.6774E+03	0.3371E+03	-0.4083E+02
$\Xi^0\to p\mu\nu$	-0.6274E+03	0.3097E+03	-0.3724E+02
$\Sigma^+\to n\mu\nu$	-0.4840E+02	0.1804E+02	-0.1655E+01
$\Xi^0\to\Sigma^-\mu\nu$	-0.2184E-02	0.3826E-03	-0.1675E-04

Table 3(D)

Process	$g_1\lambda_{g_1}$	$(g_1\lambda_{g_2}+\lambda_{g_1}g_2)$	$g_2\lambda_{g_2}$
$n\to pe\nu$	-0.7594E-06	0.9431E-09	-0.1157E-11
$\Sigma^+\to\Lambda e\nu$	-0.1953E-02	0.1059E-03	-0.5627E-05
$\Sigma^-\to\Lambda e\nu$	-0.3919E-02	0.2340E-03	-0.1368E-04
$\Lambda\to pe\nu$	-0.8884E+00	0.1234E+00	-0.1679E-01
$\Sigma^-\to ne\nu$	-0.9640E+01	0.1813E+01	-0.3337E+00
$\Xi^-\to\Lambda e\nu$	-0.1801E+01	0.2451E+00	-0.3265E-01
$\Xi^-\to\Sigma^0 e\nu$	-0.7457E-01	0.6359E-02	-0.5311E-03
$\Sigma^0\to pe\nu$	-0.8843E+01	0.1647E+01	-0.3001E+00
$\Xi^0\to\Sigma^+ e\nu$	-0.6292E-01	0.5253E-02	-0.4295E-03
$\Xi^-\to\Xi^0 e\nu$	-0.6282E-10	0.2675E-12	-0.1116E-14
$\Sigma^-\to\Sigma^0 e\nu$	-0.1096E-10	0.3919E-13	-0.1374E-15
$\Sigma^0\to\Sigma^+ e\nu$	-0.4249E-12	0.9726E-15	-0.2185E-17
$\Xi^-\to ne\nu$	-0.1085E+03	0.2737E+02	-0.6753E+01
$\Xi^0\to pe\nu$	-0.1002E+03	0.2504E+02	-0.6126E+01
$\Sigma^+\to ne\nu$	-0.7910E+01	0.1451E+01	-0.2606E+00
$\Xi^0\to\Sigma^- e\nu$	-0.4009E-01	0.3134E-02	-0.2400E-03
$\Lambda\to p\mu\nu$	-0.1188E+00	0.1754E-01	-0.2574E-02
$\Sigma^-\to n\mu\nu$	-0.4033E+01	0.7838E+00	-0.1506E+00
$\Xi^-\to\Lambda\mu\nu$	-0.4311E+00	0.6152E-01	-0.8709E-02
$\Xi^-\to\Sigma^0\mu\nu$	-0.5588E-03	0.5263E-04	-0.4952E-05
$\Sigma^0\to p\mu\nu$	-0.3603E+01	0.6938E+00	-0.1321E+00
$\Xi^0\to\Sigma^+\mu\nu$	-0.2910E-03	0.2694E-04	-0.2492E-05
$\Xi^-\to n\mu\nu$	-0.7392E+02	0.1895E+02	-0.4781E+01
$\Xi^0\to p\mu\nu$	-0.6749E+02	0.1716E+02	-0.4294E+01
$\Sigma^+\to n\mu\nu$	-0.3114E+01	0.5915E+00	-0.1111E+00
$\Xi^0\to\Sigma^-\mu\nu$	-0.3345E-04	0.2932E-05	-0.2569E-06

Table 3(E)

Process	f_1g_1	f_1g_2	f_1g_3	f_2g_1	f_2g_2	f_2g_3	f_3g_1	f_3g_2	f_3g_3
$n\to pe\nu$	0.2772E+00	-0.8852E-04	-0.4104E-07	-0.3196E-03	0.4136E-06	0.1315E-10	-0.4104E-07	0.8209E-07	-0.3023E-10
$\Sigma^+\to\Lambda e\nu$	0.4317E+00	-0.8661E-02	-0.4066E-07	-0.8674E-02	0.6261E-03	0.8880E-09	-0.4066E-07	0.8221E-07	-0.8528E-09
$\Sigma^-\to\Lambda e\nu$	0.7130E+00	-0.1570E-01	-0.6640E-07	-0.1572E-01	0.1251E-02	0.1605E-08	-0.6640E-07	0.1344E-06	-0.1535E-08
$\Lambda\to pe\nu$	0.2862E+02	-0.1395E+01	-0.3156E-05	-0.1395E+01	0.2672E+00	0.1943E-06	-0.3156E-05	0.6507E-05	-0.1740E-06
$\Sigma^-\to ne\nu$	0.1651E+03	-0.1051E+02	-0.1606E-04	-0.1051E+02	0.2791E+01	0.1423E-05	-0.1606E-04	0.3354E-04	-0.1220E-05
$\Xi^-\to\Lambda e\nu$	0.6056E+02	-0.2897E+01	-0.4756E-05	-0.2898E+01	0.5429E+00	0.2859E-06	-0.4756E-05	0.9799E-05	-0.2566E-06
$\Xi^-\to\Sigma^0 e\nu$	0.6560E+01	-0.2032E+00	-0.5063E-06	-0.2033E+00	0.2336E-01	0.1798E-07	-0.5063E-06	0.1031E-05	-0.1684E-07
$\Sigma^0\to pe\nu$	0.1547E+03	-0.9762E+01	-0.1515E-04	-0.9763E+01	0.2564E+01	0.1327E-05	-0.1515E-04	0.3163E-04	-0.1139E-05
$\Xi^0\to\Sigma^+ e\nu$	0.5781E+01	-0.1755E+00	-0.4502E-06	-0.1756E+00	0.1974E-01	0.1562E-07	-0.4502E-06	0.9161E-06	-0.1466E-07
$\Xi^-\to\Xi^0 e\nu$	0.2272E-05	-0.3582E-08	-0.1702E-12	-0.4114E-08	0.2196E-10	0.2702E-15	-0.1702E-12	0.3407E-12	-0.2892E-15
$\Sigma^-\to\Sigma^o e\nu$	0.5562E-06	-0.7249E-09	-0.5074E-13	-0.9005E-09	0.3950E-11	0.6650E-16	-0.5074E-13	0.1015E-12	-0.7432E-16
$\Sigma^0\to\Sigma^+ e\nu$	0.5109E-07	-0.4058E-10	-0.4697E-14	-0.6236E-10	0.1665E-12	0.3745E-17	-0.4697E-14	0.9397E-14	-0.4739E-17
$\Xi^-\to ne\nu$	0.1000E+04	-0.8092E+02	-0.8128E-04	-0.8092E+02	0.2987E+02	0.1056E-04	-0.8128E-04	0.1731E-03	-0.8510E-05
$\Xi^o\to pe\nu$	0.9403E+03	-0.7555E+02	-0.7710E-04	-0.7556E+02	0.2761E+02	0.9898E-05	-0.7710E-04	0.1641E-03	-0.7996E-05
$\Sigma^+\to ne\nu$	0.1427E+03	-0.8893E+01	-0.1404E-04	-0.8895E+01	0.2298E+01	0.1207E-05	-0.1404E-04	0.2929E-04	-0.1039E-05
$\Xi^o\to\Sigma^- e\nu$	0.4212E+01	-0.1201E+00	-0.3274E-06	-0.1202E+00	0.1263E-01	0.1058E-07	-0.3274E-06	0.6655E-06	-0.9966E-08
$\Lambda\to p\mu\nu$	0.2291E+01	-0.1037E-01	-0.1272E-01	-0.4905E+00	0.7349E-01	0.3613E-03	-0.1272E-01	0.2580E-01	-0.1403E-02
$\Sigma^-\to n\mu\nu$	0.4771E+02	-0.1219E+01	-0.2277E+00	-0.9482E+01	0.1942E+01	0.1343E-01	-0.2277E+00	0.4688E+00	-0.2768E-01
$\Xi^-\to\Lambda\mu\nu$	0.9415E+01	-0.1308E+00	-0.3600E-01	-0.1671E+01	0.2446E+00	0.1200E-02	-0.3600E-01	0.7319E-01	-0.3548E-02
$\Xi^-\to\Sigma^0\mu\nu$	0.2660E-01	0.1385E-03	-0.1013E-03	-0.4734E-02	0.4468E-03	0.7507E-06	-0.1013E-03	0.2033E-03	-0.8499E-05
$\Sigma^0\to p\mu\nu$	0.4330E+02	-0.1070E+01	-0.2085E+00	-0.8639E+01	0.1750E+01	0.1204E-01	-0.2085E+00	0.4290E+00	-0.2526E-01
$\Xi^0\to\Sigma^+\mu\nu$	0.1430E-01	0.8839E-04	-0.5503E-04	-0.2560E-02	0.2374E-03	0.3507E-06	-0.5503E-04	0.1104E-03	-0.4608E-05
$\Xi^-\to n\mu\nu$	0.5308E+03	-0.2629E+02	-0.2058E+01	-0.9773E+02	0.2830E+02	0.2150E+00	-0.2058E+01	0.4332E+01	-0.2881E+00
$\Xi^o\to p\mu\nu$	0.4917E+03	-0.2395E+02	-0.1927E+01	-0.9085E+02	0.2602E+02	0.1979E+00	-0.1927E+01	0.4051E+01	-0.2684E+00
$\Sigma^+\to n\mu\nu$	0.3838E+02	-0.9091E+00	-0.1858E+00	-0.7675E+01	0.1529E+01	0.1041E-01	-0.1858E+00	0.3819E+00	-0.2237E-01
$\Xi^o\to\Sigma^-\mu\nu$	0.1829E-02	0.1520E-04	-0.7036E-05	-0.3280E-03	0.2877E-04	0.2674E-07	-0.7036E-05	0.1410E-03	-0.5795E-06

Table 3(F)

Process	$(f_1\lambda_{g_1}+\lambda_{f_1}g_1)$	$(f_1\lambda_{g_2}+\lambda_{f_1}g_2)$	$(f_1\lambda_{g_3}+\lambda_{f_1}g_3)$	$(f_2\lambda_{g_1}+\lambda_{f_2}g_1)$	$(f_2\lambda_{g_2}+\lambda_{f_2}g_2)$	$(f_2\lambda_{g_3}+\lambda_{f_2}g_3)$	$(f_3\lambda_{g_1}+\lambda_{f_3}g_1)$	$(f_3\lambda_{g_2}+\lambda_{f_3}g_2)$	$(f_3\lambda_{g_3}+\lambda_{f_3}g_3)$
$n\to pe\nu$	0.2305E-06	-0.7583E-10	-0.3806E-13	-0.3281E-09	0.4241E-12	0.1516E-16	-0.2632E-10	0.1123E-12	-0.3186E-16
$\Sigma^+\to\Lambda e\nu$	0.4694E-03	-0.1423E-04	-0.3652E-10	-0.1425E-04	0.9966E-06	0.1462E-11	-0.1741E-08	0.1802E-09	-0.1401E-11
$\Sigma^-\to\Lambda e\nu$	0.9382E-03	-0.3125E-04	-0.7210E-10	-0.3127E-04	0.2412E-05	0.3202E-11	-0.3141E-08	0.3569E-09	-0.3056E-11
$\Lambda\to pe\nu$	0.2004E+00	-0.1497E-01	-0.1790E-07	-0.1497E-01	0.2761E-02	0.2095E-08	-0.3659E-06	0.9320E-07	-0.1867E-08
$\Sigma^-\to ne\nu$	0.2093E+01	-0.2061E+00	-0.1623E-06	-0.2061E+00	0.5248E-01	0.2810E-07	-0.2602E-05	0.8780E-06	-0.2392E-07
$\Xi^-\to\Lambda e\nu$	0.4071E+00	-0.2983E-01	-0.2591E-07	-0.2984E-01	0.5385E-02	0.2958E-08	-0.5391E-06	0.1347E-06	-0.2642E-08
$\Xi^-\to\Sigma^0 e\nu$	0.1752E-01	-0.8241E-03	-0.1109E-08	-0.8243E-03	0.9158E-04	0.7313E-10	-0.3479E-07	0.5576E-08	-0.6830E-10
$\Sigma^0\to pe\nu$	0.1923E+01	-0.1876E+00	-0.1503E-06	-0.1877E+00	0.4728E-01	0.2568E-07	-0.2429E-05	0.8120E-06	-0.2190E-07
$\Xi^0\to\Sigma^+ e\nu$	0.1480E-01	-0.6822E-03	-0.9462E-09	-0.6824E-03	0.7418E-04	0.6090E-10	-0.3025E-07	0.4751E-08	-0.5696E-10
$\Xi^-\to\Xi^0 e\nu$	0.1583E-10	-0.3702E-13	-0.1039E-17	-0.4087E-13	0.2158E-15	0.2832E-20	-0.5397E-15	0.4698E-17	-0.2947E-20
$\Sigma^-\to\Sigma^0 e\nu$	0.2783E-11	-0.5325E-14	-0.2286E-18	-0.6285E-14	0.2745E-16	0.5006E-21	-0.1329E-15	0.9983E-18	-0.5369E-21
$\Sigma^0\to\Sigma^+ e\nu$	0.1107E-12	-0.1241E-15	-0.9780E-20	-0.1804E-15	0.4834E-18	0.1214E-22	-0.7487E-17	0.3901E-19	-0.1436E-22
$\Xi^-\to ne\nu$	0.2240E+02	-0.2839E+01	-0.1413E-05	-0.2839E+01	0.9991E+00	0.3745E-06	-0.1843E-04	0.8117E-05	-0.2986E-06
$\Xi^o\to pe\nu$	0.2071E+02	-0.2606E+01	-0.1319E-05	-0.2606E+01	0.9082E+00	0.3451E-06	-0.1731E-04	0.7564E-05	-0.2758E-06
$\Sigma^+\to ne\nu$	0.1724E+01	-0.1660E+00	-0.1355E-06	-0.1660E+00	0.4116E-01	0.2269E-07	-0.2214E-05	0.7301E-06	-0.1939E-07
$\Xi^0\to\Sigma^- e\nu$	0.9468E-02	-0.4097E-03	-0.6049E-09	-0.4098E-03	0.4164E-04	0.3616E-10	-0.2053E-07	0.3027E-08	-0.3398E-10
$\Lambda\to p\mu\nu$	0.3263E-01	0.3807E-03	-0.2112E-03	-0.8438E-02	0.1259E-02	0.6693E-05	-0.8645E-03	0.5329E-03	-0.2435E-04
$\Sigma^-\to n\mu\nu$	0.9353E+00	-0.1646E-01	-0.5164E-02	-0.2389E+00	0.4865E-01	0.3828E-03	-0.2835E-01	0.1570E-01	-0.7168E-03
$\Xi^-\to\Lambda\mu\nu$	0.1143E+00	-0.2746E-03	-0.5102E-03	-0.2509E-01	0.3655E-02	0.1980E-04	-0.2682E-02	0.1378E-02	-0.5407E-04
$\Xi^-\to\Sigma^0\mu\nu$	0.1991E-03	0.2496E-05	-0.8209E-06	-0.3857E-04	0.3636E-05	0.6286E-08	-0.2241E-05	0.1787E-05	-0.6932E-07
$\Sigma^0\to p\mu\nu$	0.8396E+00	-0.1359E-01	-0.4685E-02	-0.2150E+00	0.4330E-01	0.3382E-03	-0.2549E-01	0.1414E-01	-0.6455E-03
$\Xi^0\to\Sigma^+\mu\nu$	0.1057E-03	0.1348E-05	-0.4365E-06	-0.2039E-04	0.1889E-05	0.2860E-08	-0.1101E-05	0.9394E-06	-0.3673E-07
$\Xi^-\to n\mu\nu$	0.1525E+02	-0.9459E+00	-0.6315E-01	-0.3738E+01	0.1079E+01	0.9590E-02	-0.4109E+00	0.2364E+00	-0.1160E-01
$\Xi^o\to p\mu\nu$	0.1397E+02	-0.8434E+00	-0.5863E-01	-0.3431E+01	0.9797E+00	0.8713E-02	-0.3795E+00	0.2179E+00	-0.1066E-01
$\Sigma^+\to n\mu\nu$	0.7308E+00	-0.1060E-01	-0.4108E-02	-0.1872E+00	0.3709E-01	0.2859E-03	-0.2214E-01	0.1229E-01	-0.5595E-03
$\Xi^0\to\Sigma^-\mu\nu$	0.1281E-04	0.1617E-06	-0.5162E-07	-0.2410E-05	0.2113E-06	0.1993E-09	-0.1025E-06	0.1080E-06	-0.4259E-08

Table 4(A)

Process	f_1^2	$f_1 f_2$	f_2^2
$n \to pe\nu$	0.6465E-04	0.1292E-03	0.6140E-07
$\Sigma^+ \to \Lambda e\nu$	0.4812E-02	0.9414E-02	0.1720E-03
$\Sigma^- \to \Lambda e\nu$	0.8817E-02	0.1721E-01	0.3471E-03
$\Lambda \to pe\nu$	0.9262E+00	0.1752E+01	0.8569E-01
$\Sigma^- \to ne\nu$	0.7815E+01	0.1450E+02	0.9834E+00
$\Xi^- \to \Lambda e\nu$	0.1911E+01	0.3619E+01	0.1732E+00
$\Xi^- \to \Sigma^0 e\nu$	0.1202E+00	0.2323E+00	0.6784E-02
$\Sigma^0 \to pe\nu$	0.7225E+01	0.1341E+02	0.9001E+00
$\Xi^0 \to \Sigma^+ e\nu$	0.1035E+00	0.2000E+00	0.5713E-02
$\Xi^- \to \Xi^0 e\nu$	0.1868E-08	0.3729E-08	0.5257E-11
$\Sigma^- \to \Sigma^0 e\nu$	0.3848E-09	0.7685E-09	0.9164E-12
$\Sigma^0 \to \Sigma^+ e\nu$	0.2271E-10	0.4538E-10	0.3542E-13
$\Xi^- \to ne\nu$	0.7059E+02	0.1278E+03	0.1197E+02
$\Xi^0 \to pe\nu$	0.6554E+02	0.1187E+03	0.1101E+02
$\Sigma^+ \to ne\nu$	0.6540E+01	0.1215E+02	0.8026E+00
$\Xi^0 \to \Sigma^- e\nu$	0.7005E-01	0.1357E+00	0.3620E-02
$\Lambda \to p\mu\nu$	0.7658E-01	0.1463E+00	0.9829E-02
$\Sigma^- \to n\mu\nu$	0.2508E+01	0.4694E+01	0.3872E+00
$\Xi^- \to \Lambda\mu\nu$	0.3154E+00	0.6021E+00	0.3752E-01
$\Xi^- \to \Sigma^0 \mu\nu$	0.3661E-03	0.7137E-03	0.3289E-04
$\Sigma^0 \to p\mu\nu$	0.2246E+01	0.4205E+01	0.3445E+00
$\Xi^0 \to \Sigma^+ \mu\nu$	0.1820E-03	0.3550E-03	0.1621E-04
$\Xi^- \to n\mu\nu$	0.4163E+02	0.7586E+02	0.7931E+01
$\Xi^0 \to p\mu\nu$	0.3811E+02	0.6952E+02	0.7216E+01
$\Sigma^+ \to n\mu\nu$	0.1950E+01	0.3656E+01	0.2961E+00
$\Xi^0 \to \Sigma^- \mu\nu$	0.1770E-04	0.3462E-04	0.1525E-05

Table 4(B)

Process	$f_1\lambda_{f_1}$	$(f_1\lambda_{f_2}+\lambda_{f_1}f_2)$	$f_2\lambda_{f_2}$
$n \to pe\nu$	0.1363E-09	0.1363E-09	0.1372E-12
$\Sigma^+ \to \Lambda e\nu$	0.1584E-04	0.1552E-04	0.6613E-06
$\Sigma^- \to \Lambda e\nu$	0.3516E-04	0.3438E-04	0.1617E-05
$\Lambda \to pe\nu$	0.1997E-01	0.1897E-01	0.2163E-02
$\Sigma^- \to ne\nu$	0.3086E+00	0.2881E+00	0.4552E-01
$\Xi^- \to \Lambda e\nu$	0.3955E-01	0.3761E-01	0.4195E-02
$\Xi^- \to \Sigma^0 e\nu$	0.9779E-03	0.9471E-03	0.6451E-04
$\Sigma^0 \to pe\nu$	0.2797E+00	0.2613E+00	0.4085E-01
$\Xi^0 \to \Sigma^+ e\nu$	0.8064E-03	0.7816E-03	0.5207E-04
$\Xi^- \to \Xi^0 e\nu$	0.3842E-13	0.3836E-13	0.1251E-15
$\Sigma^- \to \Sigma^0 e\nu$	0.5634E-14	0.5626E-14	0.1543E-16
$\Sigma^0 \to \Sigma^+ e\nu$	0.1402E-15	0.1400E-15	0.2474E-18
$\Xi^- \to ne\nu$	0.5007E+01	0.4575E+01	0.9967E+00
$\Xi^0 \to pe\nu$	0.4570E+01	0.4178E+01	0.9016E+00
$\Sigma^+ \to ne\nu$	0.2458E+00	0.2299E+00	0.3536E-01
$\Xi^0 \to \Sigma^- e\nu$	0.4789E-03	0.4651E-03	0.2893E-04
$\Lambda \to p\mu\nu$	0.2623E-02	0.2508E-02	0.3444E-03
$\Sigma^- \to n\mu\nu$	0.1309E+00	0.1228E+00	0.2139E-01
$\Xi^- \to \Lambda\mu\nu$	0.9548E-02	0.9124E-02	0.1177E-02
$\Xi^- \to \Sigma^0 \mu\nu$	0.5885E-05	0.5737E-05	0.5306E-06
$\Sigma^0 \to p\mu\nu$	0.1155E+00	0.1085E+00	0.1873E-01
$\Xi^0 \to \Sigma^+ \mu\nu$	0.2862E-05	0.2793E-05	0.2556E-06
$\Xi^- \to n\mu\nu$	0.3446E+01	0.3159E+01	0.7211E+00
$\Xi^0 \to p\mu\nu$	0.3111E+01	0.2854E+01	0.6463E+00
$\Sigma^+ \to n\mu\nu$	0.9808E-01	0.9218E-01	0.1572E-01
$\Xi^0 \to \Sigma^- \mu\nu$	0.2577E-06	0.2520E-06	0.2223E-07

Table 4(C)

Process	g_1^2	g_1g_2	g_2^2
$n \to pe\nu$	0.3795E+00	-0.9159E-03	0.5415E-06
$\Sigma^+ \to \Lambda e\nu$	0.4455E+00	-0.4601E-01	0.1140E-02
$\Sigma^- \to \Lambda e\nu$	0.7380E+00	-0.8391E-01	0.2289E-02
$\Lambda \to pe\nu$	0.3101E+02	-0.8201E+01	0.5199E+00
$\Sigma^- \to ne\nu$	0.1841E+03	-0.6588E+02	0.5649E+01
$\Xi^- \to \Lambda e\nu$	0.6552E+02	-0.1697E+02	0.1054E+01
$\Xi^- \to \Sigma^o e\nu$	0.6891E+01	-0.1119E+01	0.4358E-01
$\Sigma^o \to pe\nu$	0.1723E+03	-0.6103E+02	0.5182E+01
$\Xi^o \to \Sigma^+ e\nu$	0.6066E+01	-0.9643E+00	0.3677E-01
$\Xi^- \to \Xi^o e\nu$	0.2333E-05	-0.1895E-07	0.3699E-10
$\Sigma^- \to \Sigma^o e\nu$	0.5795E-06	-0.3956E-08	0.6498E-11
$\Sigma^o \to \Sigma^+ e\nu$	0.5556E-07	-0.2435E-09	0.2576E-12
$\Xi^- \to ne\nu$	0.1157E+04	-0.5550E+03	0.6373E+02
$\Xi^o \to pe\nu$	0.1087E+04	-0.5165E+03	0.5881E+02
$\Sigma^+ \to ne\nu$	0.1587E+03	-0.5539E+02	0.4635E+01
$\Xi^o \to \Sigma^- e\nu$	0.4407E+01	-0.6560E+00	0.2343E-01
$\Lambda \to p\mu\nu$	0.4944E+01	-0.1436E+01	0.1033E+00
$\Sigma^- \to n\mu\nu$	0.8100E+02	-0.3062E+02	0.2837E+01
$\Xi^- \to \Lambda\mu\nu$	0.1759E+02	-0.4916E+01	0.3390E+00
$\Xi^- \to \Sigma^o\mu\nu$	0.8832E-01	-0.1655E-01	0.7737E-03
$\Sigma^o \to p\mu\nu$	0.7400E+02	-0.2773E+02	0.2548E+01
$\Xi^o \to \Sigma^+\mu\nu$	0.5095E-01	-0.9390E-02	0.4321E-03
$\Xi^- \to n\mu\nu$	0.7946E+03	-0.3925E+03	0.4708E+02
$\Xi^o \to p\mu\nu$	0.7384E+03	-0.3619E+03	0.4308E+02
$\Sigma^+ \to n\mu\nu$	0.6611E+02	-0.2444E+02	0.2217E+01
$\Xi^o \to \Sigma^-\mu\nu$	0.8273E-02	-0.1446E-02	0.6318E-04

Table 4(D)

Process	$g_1\lambda_{g_1}$	$(g_1\lambda_{g_2}+\lambda_{g_1}g_2)$	$g_2\lambda_{g_2}$
$n \to pe\nu$	0.9604E-06	-0.1186E-08	0.1445E-11
$\Sigma^+ \to \Lambda e\nu$	0.1953E-02	-0.1059E-03	0.5627E-05
$\Sigma^- \to \Lambda e\nu$	0.3919E-02	-0.2340E-03	0.1368E-04
$\Lambda \to pe\nu$	0.8884E+00	-0.1234E+00	0.1679E-01
$\Sigma^- \to ne\nu$	0.9641E+01	-0.1813E+01	0.3337E+00
$\Xi^- \to \Lambda e\nu$	0.1801E+01	-0.2451E+00	0.3265E-01
$\Xi^- \to \Sigma^o e\nu$	0.7457E-01	-0.6359E-02	0.5311E-03
$\Sigma^o \to pe\nu$	0.8843E+01	-0.1647E+01	0.3001E+00
$\Xi^o \to \Sigma^+ e\nu$	0.6292E-01	-0.5253E-02	0.4295E-03
$\Xi^- \to \Xi^o e\nu$	0.6351E-10	-0.2703E-12	0.1127E-14
$\Sigma^- \to \Sigma^o e\nu$	0.1117E-10	-0.3990E-13	0.1397E-15
$\Sigma^o \to \Sigma^+ e\nu$	0.4445E-12	-0.1016E-14	0.2277E-17
$\Xi^- \to ne\nu$	0.1085E+03	-0.2737E+02	0.6753E+01
$\Xi^o \to pe\nu$	0.1002E+03	-0.2504E+02	0.6126E+01
$\Sigma^+ \to ne\nu$	0.7910E+01	-0.1451E+01	0.2606E+00
$\Xi^o \to \Sigma^- e\nu$	0.4009E-01	-0.3135E-02	0.2400E-03
$\Lambda \to p\mu\nu$	0.1896E+00	-0.2782E-01	0.4053E-02
$\Sigma^- \to n\mu\nu$	0.5031E+01	-0.9723E+00	0.1855E+00
$\Xi^- \to \Lambda\mu\nu$	0.6124E+00	-0.8687E-01	0.1220E-01
$\Xi^- \to \Sigma^o\mu\nu$	0.1488E-02	-0.1396E-03	0.1308E-04
$\Sigma^o \to p\mu\nu$	0.4522E+01	-0.8660E+00	0.1637E+00
$\Xi^o \to \Sigma^+\mu\nu$	0.8347E-03	-0.7704E-04	0.7101E-05
$\Xi^- \to n\mu\nu$	0.8175E+02	-0.2088E+02	0.5242E+01
$\Xi^o \to p\mu\nu$	0.7484E+02	-0.1896E+02	0.4721E+01
$\Sigma^+ \to n\mu\nu$	0.3941E+01	-0.7442E+00	0.1388E+00
$\Xi^o \to \Sigma^-\mu\nu$	0.1236E-03	-0.1081E-04	0.9447E-06

Table 4(E)

Process	f_1g_1	f_1g_2	f_1g_3	f_2g_1	f_2g_2	f_2g_3	f_3g_1	f_3g_2	f_3g_3
$n \to pe\nu$	0.3794E+00	-0.2640E-03	-0.3681E-07	-0.1290E-03	0.2451E-06	-0.1912E-10	0.1350E-03	-0.1123E-06	-0.1911E-10
$\Sigma^+ \to \Lambda e\nu$	0.4318E+00	-0.8667E-02	-0.6292E-07	-0.8664E-02	0.6258E-03	-0.8883E-09	0.3278E-05	-0.8223E-07	-0.8525E-09
$\Sigma^- \to \Lambda e\nu$	0.7131E+00	-0.1571E-01	-0.1031E-06	-0.1570E-01	0.1251E-02	-0.1606E-08	0.4863E-05	-0.1344E-06	-0.1535E-08
$\Lambda \to pe\nu$	0.2862E+02	-0.1395E+01	-0.5186E-05	-0.1395E+01	0.2672E+00	-0.1943E-06	0.9978E-04	-0.6507E-05	-0.1740E-06
$\Sigma^- \to ne\nu$	0.1651E+03	-0.1051E+02	-0.2744E-04	-0.1051E+02	0.2791E+01	-0.1423E-05	0.3766E-03	-0.3354E-04	-0.1220E-05
$\Xi^- \to \Lambda e\nu$	0.6056E+02	-0.2897E+01	-0.7798E-05	-0.2897E+01	0.5428E+00	-0.2859E-06	0.1535E-03	-0.9799E-05	-0.2566E-06
$\Xi^- \to \Sigma^o e\nu$	0.6561E+01	-0.2033E+00	-0.8003E-06	-0.2032E+00	0.2335E-01	-0.1798E-07	0.2600E-04	-0.1031E-05	-0.1684E-07
$\Sigma^o \to pe\nu$	0.1547E+03	-0.9763E+01	-0.2586E-04	-0.9762E+01	0.2564E+01	-0.1327E-05	0.3589E-03	-0.3163E-04	-0.1139E-05
$\Xi^o \to \Sigma^+ e\nu$	0.5781E+01	-0.1755E+00	-0.7108E-06	-0.1755E+00	0.1973E-01	-0.1563E-07	0.2361E-04	-0.9162E-06	-0.1465E-07
$\Xi^- \to \Xi^o e\nu$	0.2328E-05	-0.3875E-08	-0.2378E-12	-0.3705E-08	0.2087E-10	-0.2793E-15	0.1694E-09	-0.3490E-12	-0.2784E-15
$\Sigma^- \to \Sigma^o e\nu$	0.5783E-06	-0.8239E-09	-0.6859E-13	-0.7644E-09	0.3643E-11	-0.7009E-16	0.5947E-10	-0.1056E-12	-0.6991E-16
$\Sigma^o \to \Sigma^+ e\nu$	0.5549E-07	-0.5364E-10	-0.5845E-14	-0.4522E-10	0.1411E-12	-0.4171E-17	0.8414E-11	-0.1020E-13	-0.4164E-17
$\Xi^- \to ne\nu$	0.1000E+04	-0.8092E+02	-0.1473E-03	-0.8092E+02	0.2986E+02	-0.1056E-04	0.1435E-02	-0.1731E-03	-0.8510E-05
$\Xi^o \to pe\nu$	0.9403E+03	-0.7555E+02	-0.1394E-03	-0.7555E+02	0.2761E+02	-0.9898E-05	0.1372E-02	-0.1641E-03	-0.7995E-05
$\Sigma^+ \to ne\nu$	0.1427E+03	-0.8894E+01	-0.2391E-04	-0.8894E+01	0.2298E+01	-0.1207E-05	0.3375E-03	-0.2929E-04	-0.1039E-05
$\Xi^o \to \Sigma^- e\nu$	0.4213E+01	-0.1202E+00	-0.5151E-06	-0.1202E+00	0.1262E-01	-0.1058E-07	0.1833E-04	-0.6655E-06	-0.9964E-08
$\Lambda \to p\mu\nu$	0.4749E+01	-0.4896E+00	-0.8704E-02	-0.1054E+00	0.3003E-01	-0.6869E-03	0.3755E+00	-0.4435E-01	-0.6084E-03
$\Sigma^- \to n\mu\nu$	0.7499E+02	-0.8000E+01	-0.2192E+00	-0.3124E+01	0.1082E+01	-0.1953E-01	0.4657E+01	-0.6308E+00	-0.1659E-01
$\Xi^- \to \Lambda\mu\nu$	0.1678E+02	-0.1520E+01	-0.2844E-01	-0.4456E+00	0.1158E+00	-0.2017E-02	0.1046E+01	-0.1125E+00	-0.1795E-02
$\Xi^- \to \Sigma^o \mu\nu$	0.8732E-01	-0.7165E-02	-0.4028E-04	-0.5726E-03	0.1110E-03	-0.2341E-05	0.6553E-02	-0.5647E-03	-0.2171E-05
$\Sigma^o \to p\mu\nu$	0.6860E+02	-0.7314E+01	-0.1986E+00	-0.2807E+01	0.9660E+00	-0.1763E-01	0.4308E+01	-0.5810E+00	-0.1499E-01
$\Xi^o \to \Sigma^+ \mu\nu$	0.5045E-01	-0.4144E-02	-0.2030E-04	-0.2855E-03	0.5487E-04	-0.1175E-05	0.3838E-02	-0.3290E-03	-0.1091E-05
$\Xi^- \to n\mu\nu$	0.7030E+03	-0.7921E+02	-0.2565E+01	-0.4479E+02	0.1957E+02	-0.2662E+00	0.3186E+02	-0.5081E+01	-0.2129E+00
$\Xi^o \to p\mu\nu$	0.6543E+03	-0.7364E+02	-0.2382E+01	-0.4122E+02	0.1788E+02	-0.2461E+00	0.3004E+02	-0.4768E+01	-0.1973E+00
$\Sigma^+ \to n\mu\nu$	0.6141E+02	-0.6526E+01	-0.1747E+00	-0.2452E+01	0.8346E+00	-0.1539E-01	0.3900E+01	-0.5217E+00	-0.1312E-01
$\Xi^o \to \Sigma^- \mu\nu$	0.8224E-02	-0.6694E-03	-0.2020E-05	-0.2805E-04	0.5211E-05	-0.1143E-06	0.6394E-03	-0.5342E-04	-0.1066E-06

Table 4(F)

Process	$(f_1\lambda_{g_1}+\lambda_{f_1}g_1)$	$(f_1\lambda_{g_2}+\lambda_{f_1}g_2)$	$(f_1\lambda_{g_3}+\lambda_{f_1}g_3)$	$(f_2\lambda_{g_1}+\lambda_{f_2}g_1)$	$(f_2\lambda_{g_2}+\lambda_{f_2}g_2)$	$(f_2\lambda_{g_3}+\lambda_{f_2}g_3)$	$(f_3\lambda_{g_1}+\lambda_{f_3}g_1)$	$(f_3\lambda_{g_2}+\lambda_{f_3}g_2)$	$(f_3\lambda_{g_3}+\lambda_{f_3}g_3)$
$n\to pe\nu$	0.3696E-06	-0.2905E-09	-0.3539E-13	-0.1360E-09	0.2512E-12	-0.2016E-16	0.1545E-09	-0.1420E-12	-0.2014E-16
$\Sigma^+\to\Lambda e\nu$	0.4695E-03	-0.1424E-04	-0.6925E-10	-0.1424E-04	0.9964E-06	-0.1462E-11	0.5047E-08	-0.1802E-09	-0.1401E-11
$\Sigma^-\to\Lambda e\nu$	0.9385E-03	-0.3126E-04	-0.1375E-09	-0.3126E-04	0.2412E-05	-0.3203E-11	0.9072E-08	-0.3569E-09	-0.3055E-11
$\Lambda\to pe\nu$	0.2004E+00	-0.1497E-01	-0.3742E-07	-0.1497E-01	0.2761E-02	-0.2095E-08	0.1007E-05	-0.9320E-07	-0.1867E-08
$\Sigma^-\to ne\nu$	0.2093E+01	-0.2061E+00	-0.3630E-06	-0.2061E+00	0.5248E-01	-0.2810E-07	0.6957E-05	-0.8780E-06	-0.2392E-07
$\Xi^-\to\Lambda e\nu$	0.4072E+00	-0.2983E-01	-0.5399E-07	-0.2983E-01	0.5384E-02	-0.2958E-08	0.1486E-05	-0.1347E-06	-0.2642E-08
$\Xi^-\to\Sigma^o e\nu$	0.1752E-01	-0.8243E-03	-0.2176E-08	-0.8242E-03	0.9157E-04	-0.7313E-10	0.9887E-07	-0.5576E-08	-0.6829E-10
$\Sigma^o\to pe\nu$	0.1923E+01	-0.1877E+00	-0.3353E-06	-0.1877E+00	0.4728E-01	-0.2568E-07	0.6500E-05	-0.8120E-06	-0.2190E-07
$\Xi^o\to\Sigma^+ e\nu$	0.1480E-01	-0.6823E-03	-0.1852E-08	-0.6823E-03	0.7417E-04	-0.6090E-10	0.8607E-07	-0.4751E-08	-0.5696E-10
$\Xi^-\to\Xi^o e\nu$	0.1648E-10	-0.3980E-13	-0.1747E-17	-0.3811E-13	0.2081E-15	-0.2873E-20	0.1694E-14	-0.4749E-17	-0.2864E-20
$\Sigma^-\to\Sigma^o e\nu$	0.2975E-11	-0.6025E-14	-0.3685E-18	-0.5595E-14	0.2582E-16	-0.5131E-21	0.4299E-15	-0.1017E-17	-0.5117E-21
$\Sigma^o\to\Sigma^+ e\nu$	0.1277E-12	-0.1661E-15	-0.1408E-19	-0.1395E-15	0.4206E-18	-0.1287E-22	0.2651E-16	-0.4082E-19	-0.1285E-22
$\Xi^-\to ne\nu$	0.2240E+02	-0.2839E+01	-0.3505E-05	-0.2839E+01	0.9991E+00	-0.3745E-06	0.4756E-04	-0.8117E-05	-0.2986E-06
$\Xi^o\to pe\nu$	0.2071E+02	-0.2606E+01	-0.3261E-05	-0.2606E+01	0.9081E+00	-0.3451E-06	0.4471E-04	-0.7564E-05	-0.2758E-06
$\Sigma^+\to ne\nu$	0.1724E+01	-0.1660E+00	-0.3010E-06	-0.1660E+00	0.4116E-01	-0.2269E-07	0.5934E-05	-0.7301E-06	-0.1939E-07
$\Xi^o\to\Sigma^- e\nu$	0.9469E-02	-0.4098E-03	-0.1177E-08	-0.4097E-03	0.4164E-04	-0.3617E-10	0.5861E-07	-0.3028E-08	-0.3398E-10
$\Lambda\to p\mu\nu$	0.8244E-01	-0.8904E-02	-0.1446E-03	-0.1808E-02	0.5060E-03	-0.1177E-04	0.6952E-02	-0.8506E-03	-0.1041E-04
$\Sigma^-\to n\mu\nu$	0.1821E+01	-0.2192E+00	-0.5260E-02	-0.8169E-01	0.2722E-01	-0.5096E-03	0.1322E+00	-0.1959E-01	-0.4314E-03
$\Xi^-\to\Lambda\mu\nu$	0.2504E+00	-0.2440E-01	-0.4096E-03	-0.6755E-02	0.1712E-02	-0.3053E-04	0.1723E-01	-0.1958E-02	-0.2713E-04
$\Xi^-\to\Sigma^o\mu\nu$	0.7230E-03	-0.5979E-04	-0.3223E-06	-0.4604E-05	0.8899E-06	-0.1882E-07	0.5486E-04	-0.4756E-05	-0.1745E-07
$\Sigma^o\to p\mu\nu$	0.1649E+01	-0.1977E+00	-0.4710E-02	-0.7238E-01	0.2398E-01	-0.4535E-03	0.1206E+00	-0.1775E-01	-0.3846E-03
$\Xi^o\to\Sigma^+\mu\nu$	0.4080E-03	-0.3371E-04	-0.1591E-06	-0.2246E-05	0.4306E-06	-0.9241E-08	0.3130E-04	-0.2695E-05	-0.8583E-08
$\Xi^-\to n\mu\nu$	0.2399E+02	-0.3353E+01	-0.9115E-01	-0.1855E+01	0.7668E+00	-0.1102E-01	0.1406E+01	-0.2614E+00	-0.8751E-02
$\Xi^o\to p\mu\nu$	0.2212E+02	-0.3077E+01	-0.8369E-01	-0.1684E+01	0.6914E+00	-0.1004E-01	0.1309E+01	-0.2416E+00	-0.7997E-02
$\Sigma^+\to n\mu\nu$	0.1450E+01	-0.1727E+00	-0.4061E-02	-0.6182E-01	0.2027E-01	-0.3870E-03	0.1068E+00	-0.1555E-01	-0.3291E-03
$\Xi^o\to\Sigma^-\mu\nu$	0.6110E-04	-0.4985E-05	-0.1469E-07	-0.2042E-06	0.3790E-07	-0.8320E-09	0.4766E-05	-0.3989E-06	-0.7761E-09

Table 5(A)

Process	f_1g_1	f_1g_2	f_1g_3	f_2g_1	f_2g_2	f_3g_1	f_3g_3
$n\to pe\nu$	-0.4170E+00	0.2184E-03	0.5624E-07	0.4372E-03	-0.6019E-06	-0.8165E-04	0.4250E-10
$\Sigma^+\to\Lambda e\nu$	-0.5395E+00	0.1067E-01	0.7839E-07	0.2243E-01	-0.1402E-02	-0.2450E-05	0.1628E-08
$\Sigma^-\to\Lambda e\nu$	-0.8909E+00	0.1930E-01	0.1285E-06	0.4079E-01	-0.2810E-02	-0.3635E-05	0.2936E-08
$\Lambda\to pe\nu$	-0.3570E+02	0.1658E+01	0.6451E-05	0.3836E+01	-0.6235E+00	-0.7472E-04	0.3426E-06
$\Sigma^-\to ne\nu$	-0.2055E+03	0.1214E+02	0.3409E-04	0.3007E+02	-0.6681E+01	-0.2826E-03	0.2449E-05
$\Xi^-\to\Lambda e\nu$	-0.7554E+02	0.3449E+01	0.9700E-05	0.7948E+01	-0.1265E+01	-0.1149E-03	0.5047E-06
$\Xi^-\to\Sigma^0 e\nu$	-0.8194E+01	0.2474E+00	0.9964E-06	0.5373E+00	-0.5309E-01	-0.1944E-04	0.3251E-07
$\Sigma^0\to pe\nu$	-0.1925E+03	0.1129E+02	0.3212E-04	0.2788E+02	-0.6131E+01	-0.2693E-03	0.2285E-05
$\Xi^0\to\Sigma^+ e\nu$	-0.7220E+01	0.2138E+00	0.8849E-06	0.4634E+00	-0.4481E-01	-0.1765E-04	0.2827E-07
$\Xi^-\to\Xi^0 e\nu$	-0.2892E-05	0.4710E-08	0.3031E-12	0.9454E-08	-0.4596E-10	-0.1245E-09	0.5275E-15
$\Sigma^-\to\Sigma^0 e\nu$	-0.7154E-06	0.9837E-09	0.8828E-13	0.1973E-08	-0.8047E-11	-0.4323E-10	0.1332E-15
$\Sigma^0\to\Sigma^+ e\nu$	-0.6767E-07	0.6051E-10	0.7731E-14	0.1213E-09	-0.3153E-12	-0.5940E-11	0.8094E-17
$\Xi^-\to ne\nu$	-0.1240E+04	0.8865E+02	0.1825E-03	0.2451E+03	-0.7409E+02	-0.1081E-02	0.1755E-04
$\Xi^0\to pe\nu$	-0.1166E+04	0.8295E+02	0.1728E-03	0.2284E+03	-0.6840E+02	-0.1034E-02	0.1647E-04
$\Sigma^+\to ne\nu$	-0.1776E+03	0.1030E+02	0.2970E-04	0.2534E+02	-0.5488E+01	-0.2532E-03	0.2082E-05
$\Xi^0\to\Sigma^- e\nu$	-0.5262E+01	0.1467E+00	0.6413E-06	0.3161E+00	-0.2860E-01	-0.1371E-04	0.1919E-07
$\Lambda\to p\mu\nu$	-0.4227E+01	0.2510E+00	0.1607E-01	0.5806E+00	-0.9349E-01	-0.1862E+00	0.1764E-02
$\Sigma^-\to n\mu\nu$	-0.7851E+02	0.5411E+01	0.3374E+00	0.1340E+02	-0.2958E+01	-0.2797E+01	0.4058E-01
$\Xi^-\to\Lambda\mu\nu$	-0.1631E+02	0.9124E+00	0.4805E-01	0.2103E+01	-0.3323E+00	-0.5692E+00	0.4722E-02
$\Xi^-\to\Sigma^0\mu\nu$	-0.5386E-01	0.2296E-02	0.1107E-03	0.4984E-02	-0.4879E-03	-0.2159E-02	0.9267E-05
$\Sigma^0\to p\mu\nu$	-0.7153E+02	0.4909E+01	0.3072E+00	0.1212E+02	-0.2648E+01	-0.2575E+01	0.3683E-01
$\Xi^o\to\Sigma^+\mu\nu$	-0.2925E-01	0.1234E-02	0.5944E-04	0.2675E-02	-0.2562E-03	-0.1186E-02	0.4967E-05
$\Xi^-\to n\mu\nu$	-0.7943E+03	0.6215E+02	0.3585E+01	0.1718E+03	-0.5170E+02	-0.2123E+02	0.4864E+00
$\Xi^0\to p\mu\nu$	-0.7380E+03	0.5757E+02	0.3337E+01	0.1585E+03	-0.4725E+02	-0.1997E+02	0.4512E+00
$\Sigma^+\to n\mu\nu$	-0.6370E+02	0.4339E+01	0.2717E+00	0.1067E+02	-0.2295E+01	-0.2316E+01	0.3239E-01
$\Xi^0\to\Sigma^-\mu\nu$	-0.3838E-02	0.1562E-03	0.7382E-05	0.3366E-03	-0.3015E-04	-0.1577E-03	0.6071E-06

Table 5(B)

Process	$(f_1\lambda_{g_1}+\lambda_{f_1}g_1)$	$(f_1\lambda_{g_2}+\lambda_{f_1}g_2)$	$(f_1\lambda_{g_3}+\lambda_{f_1}g_3)$	$(f_2\lambda_{g_1}+\lambda_{f_2}g_1)$	$(f_2\lambda_{g_2}+\lambda_{f_2}g_2)$	$(f_3\lambda_{g_1}+\lambda_{f_3}g_1)$	$(f_3\lambda_{g_3}+\lambda_{f_3}g_3)$
$n\to pe\nu$	-0.4132E-06	0.2493E-09	0.5851E-13	0.4992E-09	-0.6873E-12	-0.8494E-10	0.4914E-16
$\Sigma^+\to\Lambda e\nu$	-0.6843E-03	0.2051E-04	0.1010E-09	0.4313E-04	-0.2692E-05	-0.3155E-08	0.3131E-11
$\Sigma^-\to\Lambda e\nu$	-0.1368E-02	0.4496E-04	0.2005E-09	0.9505E-04	-0.6535E-05	-0.5672E-08	0.6842E-11
$\Lambda\to pe\nu$	-0.2911E+00	0.2093E-01	0.5446E-07	0.4842E-01	-0.7835E-02	-0.6308E-06	0.4325E-08
$\Sigma^-\to ne\nu$	-0.3029E+01	0.2809E+00	0.5272E-06	0.6957E+00	-0.1536E+00	-0.4371E-05	0.5667E-07
$\Xi^-\to\Lambda e\nu$	-0.5916E+00	0.4177E-01	0.7858E-07	0.9627E-01	-0.1525E-01	-0.9309E-06	0.6113E-08
$\Xi^-\to\Sigma^0 e\nu$	-0.2551E-01	0.1176E-02	0.3171E-08	0.2554E-02	-0.2517E-03	-0.6185E-07	0.1546E-09
$\Sigma^0\to pe\nu$	-0.2783E+01	0.2560E+00	0.4871E-06	0.6324E+00	-0.1382E+00	-0.4083E-05	0.5184E-07
$\Xi^0\to\Sigma^+ e\nu$	-0.2156E-01	0.9744E-03	0.2699E-08	0.2112E-02	-0.2037E-03	-0.5384E-07	0.1288E-09
$\Xi^-\to\Xi^0 e\nu$	-0.2354E-10	0.5612E-13	0.2563E-17	0.1126E-12	-0.5476E-15	-0.1052E-14	0.6302E-20
$\Sigma^-\to\Sigma^0 e\nu$	-0.4193E-11	0.8303E-14	0.5430E-18	0.1666E-13	-0.6791E-16	-0.2659E-15	0.1129E-20
$\Sigma^0\to\Sigma^+ e\nu$	-0.1730E-12	0.2125E-15	0.2104E-19	0.4259E-15	-0.1108E-17	-0.1617E-16	0.2865E-22
$\Xi^-\to ne\nu$	-0.3216E+02	0.3689E+01	0.5070E-05	0.1020E+02	-0.3056E+01	-0.3003E-04	0.7303E-06
$\Xi^0\to pe\nu$	-0.2974E+02	0.3393E+01	0.4718E-05	0.9341E+01	-0.2773E+01	-0.2823E-04	0.6738E-06
$\Sigma^+\to ne\nu$	-0.2496E+01	0.2269E+00	0.4373E-06	0.5581E+00	-0.1201E+00	-0.3727E-05	0.4585E-07
$\Xi^0\to\Sigma^- e\nu$	-0.1379E-01	0.5861E-03	0.1715E-08	0.1263E-02	-0.1140E-03	-0.3666E-07	0.7667E-10
$\Lambda\to p\mu\nu$	-0.7189E-01	0.4514E-02	0.2804E-03	0.1044E-01	-0.1679E-02	-0.3248E-02	0.3196E-04
$\Sigma^-\to n\mu\nu$	-0.1912E+01	0.1528E+00	0.8735E-02	0.3784E+00	-0.8323E-01	-0.7242E-01	0.1160E-02
$\Xi^-\to\Lambda\mu\nu$	-0.2404E+00	0.1467E-01	0.7352E-03	0.3382E-01	-0.5335E-02	-0.8710E-02	0.7672E-04
$\Xi^-\to\Sigma^0\mu\nu$	-0.4379E-03	0.1883E-04	0.9038E-06	0.4088E-04	-0.4001E-05	-0.1763E-04	0.7614E-07
$\Sigma^o\to p\mu\nu$	-0.1722E+01	0.1365E+00	0.7851E-02	0.3371E+00	-0.7339E-01	-0.6581E-01	0.1037E-02
$\Xi^o\to\Sigma^+\mu\nu$	-0.2327E-03	0.9882E-05	0.4743E-06	0.2142E-04	-0.2051E-05	-0.9460E-05	0.3982E-07
$\Xi^-\to n\mu\nu$	-0.2819E+02	0.2863E+01	0.1405E+00	0.7917E+01	-0.2366E+01	-0.8322E+00	0.2268E-01
$\Xi^0\to p\mu\nu$	-0.2591E+02	0.2614E+01	0.1292E+00	0.7196E+01	-0.2131E+01	-0.7732E+00	0.2074E-01
$\Sigma^+\to n\mu\nu$	-0.1505E+01	0.1178E+00	0.6802E-02	0.2898E+00	-0.6213E-01	-0.5798E-01	0.8901E-03
$\Xi^0\to\Sigma^-\mu\nu$	-0.2818E-04	0.1150E-05	0.5428E-07	0.2478E-05	-0.2220E-06	-0.1160E-05	0.4472E-08

Table 6(A)

Process	f_1^2	f_1f_2	f_2^2
$n\to pe\nu$	-0.7707E-05	-0.1541E-04	-0.2121E-07
$\Sigma^+\to\Lambda e\nu$	-0.4483E-06	-0.8967E-06	-0.5388E-07
$\Sigma^-\to\Lambda e\nu$	-0.6705E-06	-0.1341E-05	-0.8843E-07
$\Lambda\to pe\nu$	-0.1513E-04	-0.3026E-04	-0.4427E-05
$\Sigma^-\to ne\nu$	-0.6055E-04	-0.1211E-03	-0.2327E-04
$\Xi^-\to\Lambda e\nu$	-0.2324E-04	-0.4647E-04	-0.6672E-05
$\Xi^-\to\Sigma^0 e\nu$	-0.3710E-05	-0.7419E-05	-0.6883E-06
$\Sigma^0\to pe\nu$	-0.5757E-04	-0.1151E-03	-0.2193E-04
$\Xi^0\to\Sigma^+ e\nu$	-0.3362E-05	-0.6723E-05	-0.6113E-06
$\Xi^-\to\Xi^0 e\nu$	-0.1778E-10	-0.3556E-10	-0.1724E-12
$\Sigma^-\to\Sigma^0 e\nu$	-0.5867E-11	-0.1173E-10	-0.4773E-13
$\Sigma^0\to\Sigma^+ e\nu$	-0.7272E-12	-0.1454E-11	-0.3776E-14
$\Xi^-\to ne\nu$	-0.2498E-03	-0.4997E-03	-0.1235E-03
$\Xi^0\to pe\nu$	-0.2383E-03	-0.4766E-03	-0.1170E-03
$\Sigma^+\to ne\nu$	-0.5396E-04	-0.1079E-03	-0.2028E-04
$\Xi^0\to\Sigma^- e\nu$	-0.2593E-05	-0.5187E-05	-0.4430E-06
$\Lambda\to p\mu\nu$	-0.1652E-01	-0.3303E-01	-0.4833E-02
$\Sigma^-\to n\mu\nu$	-0.2970E+00	-0.5941E+00	-0.1141E+00
$\Xi^-\to\Lambda\mu\nu$	-0.5348E-01	-0.1070E+00	-0.1536E-01
$\Xi^-\to\Sigma^0\mu\nu$	-0.1632E-03	-0.3263E-03	-0.3027E-04
$\Sigma^0\to p\mu\nu$	-0.2716E+00	-0.5432E+00	-0.1034E+00
$\Xi^0\to\Sigma^+\mu\nu$	-0.8857E-04	-0.1771E-03	-0.1611E-04
$\Xi^-\to n\mu\nu$	-0.2735E+01	-0.5471E+01	-0.1352E+01
$\Xi^0\to p\mu\nu$	-0.2556E+01	-0.5111E+01	-0.1254E+01
$\Sigma^+\to n\mu\nu$	-0.2422E+00	-0.4844E+00	-0.9104E-01
$\Xi^0\to\Sigma^-\mu\nu$	-0.1144E-04	-0.2289E-04	-0.1955E-05

Table 6(B)

Process	$f_1\lambda_{f_1}$	$(f_1\lambda_{f_2}+\lambda_{f_1}f_2)$	$f_2\lambda_{f_2}$
$n\to pe\nu$	-0.1592E-10	-0.1592E-10	-0.4380E-13
$\Sigma^+\to\Lambda e\nu$	-0.8181E-09	-0.8181E-09	-0.9832E-10
$\Sigma^-\to\Lambda e\nu$	-0.1480E-08	-0.1480E-08	-0.1952E-09
$\Lambda\to pe\nu$	-0.1815E-06	-0.1815E-06	-0.5311E-07
$\Sigma^-\to ne\nu$	-0.1341E-05	-0.1341E-05	-0.5152E-06
$\Xi^-\to\Lambda e\nu$	-0.2668E-06	-0.2668E-06	-0.7662E-07
$\Xi^-\to\Sigma^0 e\nu$	-0.1665E-07	-0.1665E-07	-0.3089E-08
$\Sigma^0\to pe\nu$	-0.1250E-05	-0.1250E-05	-0.4759E-06
$\Xi^0\to\Sigma^+ e\nu$	-0.1446E-07	-0.1446E-07	-0.2629E-08
$\Xi^-\to\Xi^0 e\nu$	-0.2497E-15	-0.2497E-15	-0.2421E-17
$\Sigma^-\to\Sigma^0 e\nu$	-0.6202E-16	-0.6202E-16	-0.5045E-18
$\Sigma^0\to\Sigma^+ e\nu$	-0.3605E-17	-0.3605E-17	-0.1872E-19
$\Xi^-\to ne\nu$	-0.1007E-04	-0.1007E-04	-0.4979E-05
$\Xi^0\to pe\nu$	-0.9438E-05	-0.9438E-05	-0.4633E-05
$\Sigma^+\to ne\nu$	-0.1136E-05	-0.1136E-05	-0.4272E-06
$\Xi^0\to\Sigma^- e\nu$	-0.9782E-08	-0.9782E-08	-0.1671E-08
$\Lambda\to p\mu\nu$	-0.5852E-03	-0.5852E-03	-0.1713E-03
$\Sigma^-\to n\mu\nu$	-0.1570E-01	-0.1570E-01	-0.6033E-02
$\Xi^-\to\Lambda\mu\nu$	-0.1664E-02	-0.1664E-02	-0.4777E-03
$\Xi^-\to\Sigma^0\mu\nu$	-0.2679E-05	-0.2679E-05	-0.4970E-06
$\Sigma^0\to p\mu\nu$	-0.1418E-01	-0.1418E-01	-0.5399E-02
$\Xi^0\to\Sigma^+\mu\nu$	-0.1420E-05	-0.1420E-05	-0.2582E-06
$\Xi^-\to n\mu\nu$	-0.2182E+00	-0.2182E+00	-0.1079E+00
$\Xi^0\to p\mu\nu$	-0.2015E+00	-0.2015E+00	-0.9890E-01
$\Sigma^+\to n\mu\nu$	-0.1238E-01	-0.1238E-01	-0.4654E-02
$\Xi^0\to\Sigma^-\mu\nu$	-0.1687E-06	-0.1687E-06	-0.2882E-07

Table 6(C)

Process	g_1^2	g_1g_2	g_2^2
$n \to pe\nu$	-0.7707E-05	0.1541E-04	-0.2121E-07
$\Sigma^+ \to \Lambda e\nu$	-0.4483E-06	0.8967E-06	-0.5388E-07
$\Sigma^- \to \Lambda e\nu$	-0.6705E-06	0.1341E-05	-0.8843E-07
$\Lambda \to pe\nu$	-0.1513E-04	0.3026E-04	-0.4427E-05
$\Sigma^- \to ne\nu$	-0.6055E-04	0.1211E-03	-0.2327E-04
$\Xi^- \to \Lambda e\nu$	-0.2324E-04	0.4647E-04	-0.6672E-05
$\Xi^- \to \Sigma^o e\nu$	-0.3710E-05	0.7419E-05	-0.6883E-06
$\Sigma^o \to pe\nu$	-0.5757E-04	0.1151E-03	-0.2193E-04
$\Xi^o \to \Sigma^+ e\nu$	-0.3362E-05	0.6723E-05	-0.6113E-06
$\Xi^- \to \Xi^o e\nu$	-0.1778E-10	0.3556E-10	-0.1724E-12
$\Sigma^- \to \Sigma^o e\nu$	-0.5867E-11	0.1173E-10	-0.4773E-13
$\Sigma^o \to \Sigma^+ e\nu$	-0.7272E-12	0.1454E-11	-0.3776E-14
$\Xi^- \to ne\nu$	-0.2498E-03	0.4997E-03	-0.1235E-03
$\Xi^o \to pe\nu$	-0.2383E-03	0.4766E-03	-0.1170E-03
$\Sigma^+ \to ne\nu$	-0.5396E-04	0.1079E-03	-0.2028E-04
$\Xi^o \to \Sigma^- e\nu$	-0.2593E-05	0.5187E-05	-0.4430E-06
$\Lambda \to p\mu\nu$	-0.1652E-01	0.3303E-01	-0.4833E-02
$\Sigma^- \to n\mu\nu$	-0.2970E+00	0.5941E+00	-0.1141E+00
$\Xi^- \to \Lambda\mu\nu$	-0.5348E-01	0.1070E+00	-0.1536E-01
$\Xi^- \to \Sigma^o\mu\nu$	-0.1632E-03	0.3263E-03	-0.3027E-04
$\Sigma^o \to p\mu\nu$	-0.2716E+00	0.5432E+00	-0.1034E+00
$\Xi^o \to \Sigma^+\mu\nu$	-0.8857E-04	0.1771E-03	-0.1611E-04
$\Xi^- \to n\mu\nu$	-0.2735E+01	0.5471E+01	-0.1352E+01
$\Xi^o \to p\mu\nu$	-0.2556E+01	0.5111E+01	-0.1254E+01
$\Sigma^+ \to n\mu\nu$	-0.2422E+00	0.4844E+00	-0.9104E-01
$\Xi^o \to \Sigma^-\mu\nu$	-0.1144E-04	0.2289E-04	-0.1955E-05

Table 6(D)

Process	$g_1\lambda_{g_1}$	$(g_1\lambda_{g_2}+\lambda_{g_1}g_2)$	$g_2\lambda_{g_2}$
$n \to pe\nu$	-0.1592E-10	0.1592E-10	-0.4380E-13
$\Sigma^+ \to \Lambda e\nu$	-0.8181E-09	0.8181E-09	-0.9832E-10
$\Sigma^- \to \Lambda e\nu$	-0.1480E-08	0.1480E-08	-0.1952E-09
$\Lambda \to pe\nu$	-0.1815E-06	0.1815E-06	-0.5311E-07
$\Sigma^- \to ne\nu$	-0.1341E-05	0.1341E-05	-0.5152E-06
$\Xi^- \to \Lambda e\nu$	-0.2668E-06	0.2668E-06	-0.7662E-07
$\Xi^- \to \Sigma^o e\nu$	-0.1665E-07	0.1665E-07	-0.3089E-08
$\Sigma^o \to pe\nu$	-0.1250E-05	0.1250E-05	-0.4759E-06
$\Xi^o \to \Sigma^+ e\nu$	-0.1446E-07	0.1446E-07	-0.2629E-08
$\Xi^- \to \Xi^o e\nu$	-0.2497E-15	0.2497E-15	-0.2421E-17
$\Sigma^- \to \Sigma^o e\nu$	-0.6202E-16	0.6202E-16	-0.5045E-18
$\Sigma^o \to \Sigma^+ e\nu$	-0.3605E-17	0.3605E-17	-0.1872E-19
$\Xi^- \to ne\nu$	-0.1007E-04	0.1007E-04	-0.4979E-05
$\Xi^o \to pe\nu$	-0.9438E-05	0.9438E-05	-0.4633E-05
$\Sigma^+ \to ne\nu$	-0.1136E-05	0.1136E-05	-0.4272E-06
$\Xi^o \to \Sigma^- e\nu$	-0.9782E-08	0.9782E-08	-0.1671E-08
$\Lambda \to p\mu\nu$	-0.5852E-03	0.5852E-03	-0.1713E-03
$\Sigma^- \to n\mu\nu$	-0.1570E-01	0.1570E-01	-0.6033E-02
$\Xi^- \to \Lambda\mu\nu$	-0.1664E-02	0.1664E-02	-0.4777E-03
$\Xi^- \to \Sigma^o\mu\nu$	-0.2679E-05	0.2679E-05	-0.4970E-06
$\Sigma^o \to p\mu\nu$	-0.1418E-01	0.1418E-01	-0.5399E-02
$\Xi^o \to \Sigma^+\mu\nu$	-0.1420E-05	0.1420E-05	-0.2582E-06
$\Xi^- \to n\mu\nu$	-0.2182E+00	0.2182E+00	-0.1079E+00
$\Xi^o \to p\mu\nu$	-0.2015E+00	0.2015E+00	-0.9890E-01
$\Sigma^+ \to n\mu\nu$	-0.1238E-01	0.1238E-01	-0.4654E-02
$\Xi^o \to \Sigma^-\mu\nu$	-0.1687E-06	0.1687E-06	-0.2882E-07

Table 6(E)

Process	f_1g_1	f_1g_2	f_1g_3	f_2g_1	f_2g_2	f_2g_3	f_3g_1	f_3g_2	f_3g_3
$n \to pe\nu$	0.5061E+00	-0.2976E-03	0.6198E-07	0.2976E-03	-0.5596E-06	-0.4173E-10	-0.8998E-04	-0.4173E-10	-0.4401E-10
$\Sigma^+ \to \Lambda e\nu$	0.5957E+00	-0.1509E-01	0.7013E-07	0.1509E-01	-0.1053E-02	-0.1393E-08	-0.2191E-05	-0.1393E-08	-0.1393E-08
$\Sigma^- \to \Lambda e\nu$	0.9872E+00	-0.2760E-01	0.1150E-06	0.2760E-01	-0.2116E-02	-0.2513E-08	-0.3253E-05	-0.2513E-08	-0.2513E-08
$\Lambda \to pe\nu$	0.4182E+02	-0.2808E+01	0.5796E-05	0.2808E+01	-0.4894E+00	-0.2946E-06	-0.6714E-04	-0.2946E-06	-0.2946E-06
$\Sigma^- \to ne\nu$	0.2501E+03	-0.2319E+02	0.3072E-04	0.2319E+02	-0.5392E+01	-0.2112E-05	-0.2547E-03	-0.2112E-05	-0.2112E-05
$\Xi^- \to \Lambda e\nu$	0.8831E+02	-0.5801E+01	0.8715E-05	0.5801E+01	-0.9912E+00	-0.4338E-06	-0.1032E-03	-0.4338E-06	-0.4339E-06
$\Xi^- \to \Sigma^o e\nu$	0.9238E+01	-0.3726E+00	0.8928E-06	0.3726E+00	-0.4050E-01	-0.2786E-07	-0.1742E-04	-0.2786E-07	-0.2786E-07
$\Sigma^0 \to pe\nu$	0.2339E+03	-0.2146E+02	0.2894E-04	0.2146E+02	-0.4943E+01	-0.1971E-05	-0.2426E-03	-0.1971E-05	-0.1971E-05
$\Xi^o \to \Sigma^+ e\nu$	0.8130E+01	-0.3208E+00	0.7929E-06	0.3208E+00	-0.3415E-01	-0.2423E-07	-0.1582E-04	-0.2423E-07	-0.2423E-07
$\Xi^- \to \Xi^0 e\nu$	0.3112E-05	-0.6093E-08	0.2751E-12	0.6093E-08	-0.3425E-10	-0.4530E-15	-0.1130E-09	-0.4530E-15	-0.4557E-15
$\Sigma^- \to \Sigma^o e\nu$	0.7728E-06	-0.1274E-08	0.8097E-13	0.1274E-08	-0.6056E-11	-0.1149E-15	-0.3965E-10	-0.1149E-15	-0.1160E-15
$\Sigma^o \to \Sigma^+ e\nu$	0.7409E-07	-0.7864E-10	0.7301E-14	0.7864E-10	-0.2449E-12	-0.7087E-17	-0.5610E-11	-0.7087E-17	-0.7220E-17
$\Xi^- \to ne\nu$	0.1593E+04	-0.2033E+03	0.1652E-03	0.2033E+03	-0.6217E+02	-0.1521E-04	-0.9781E-03	-0.1521E-04	-0.1521E-04
$\Xi^o \to pe\nu$	0.1495E+04	-0.1890E+03	0.1563E-03	0.1890E+03	-0.5732E+02	-0.1427E-04	-0.9353E-03	-0.1427E-04	-0.1427E-04
$\Sigma^+ \to ne\nu$	0.2154E+03	-0.1945E+02	0.2676E-04	0.1945E+02	-0.4418E+01	-0.1795E-05	-0.2281E-03	-0.1795E-05	-0.1795E-05
$\Xi^o \to \Sigma^- e\nu$	0.5903E+01	-0.2177E+00	0.5744E-06	0.2177E+00	-0.2174E-01	-0.1644E-07	-0.1228E-04	-0.1644E-07	-0.1644E-07
$\Lambda \to p\mu\nu$	0.6740E+01	-0.4899E+00	0.2122E-01	0.4899E+00	-0.1175E+00	-0.2049E-02	-0.2458E+00	-0.2049E-02	-0.2197E-02
$\Sigma^- \to n\mu\nu$	0.1113E+03	-0.1084E+02	0.3725E+00	0.1084E+02	-0.3073E+01	-0.3988E-01	-0.3088E+01	-0.3988E-01	-0.4219E-01
$\Xi^- \to \Lambda\mu\nu$	0.2395E+02	-0.1685E+01	0.5821E-01	0.1685E+01	-0.3749E+00	-0.5047E-02	-0.6895E+00	-0.5047E-02	-0.5389E-02
$\Xi^- \to \Sigma^0 \mu\nu$	0.1192E+00	-0.5452E-02	0.2183E-03	0.5452E-02	-0.9395E-03	-0.1639E-04	-0.4259E-02	-0.1639E-04	-0.1743E-04
$\Sigma^0 \to p\mu\nu$	0.1016E+03	-0.9799E+01	0.3406E+00	0.9799E+01	-0.2763E+01	-0.3634E-01	-0.2856E+01	-0.3634E-01	-0.3847E-01
$\Xi^o \to \Sigma^+ \mu\nu$	0.6869E-01	-0.3089E-02	0.1249E-03	0.3089E-02	-0.5288E-03	-0.9401E-05	-0.2492E-02	-0.9401E-05	-0.9973E-05
$\Xi^- \to n\mu\nu$	0.1104E+04	-0.1449E+03	0.3619E+01	0.1449E+03	-0.4970E+02	-0.4459E+00	-0.2144E+02	-0.4459E+00	-0.4634E+00
$\Xi^o \to p\mu\nu$	0.1025E+04	-0.1334E+03	0.3377E+01	0.1334E+03	-0.4550E+02	-0.4143E+00	-0.2020E+02	-0.4143E+00	-0.4308E+00
$\Sigma^+ \to n\mu\nu$	0.9073E+02	-0.8622E+01	0.3031E+00	0.8622E+01	-0.2409E+01	-0.3211E-01	-0.2583E+01	-0.3211E-01	-0.3402E-01
$\Xi^0 \to \Sigma^- \mu\nu$	0.1113E-01	-0.4743E-03	0.1937E-04	0.4743E-03	-0.7905E-04	-0.1457E-05	-0.4139E-03	-0.1457E-05	-0.1531E-05

Table 6(F)

Process	$(f_1\lambda_{g_1}+\lambda_{f_1}g_1)$	$(f_1\lambda_{g_2}+\lambda_{f_1}g_2)$	$(f_1\lambda_{g_3}+\lambda_{f_1}g_3)$	$(f_2\lambda_{g_1}+\lambda_{f_2}g_1)$	$(f_2\lambda_{g_2}+\lambda_{f_2}g_2)$	$(f_2\lambda_{g_3}+\lambda_{f_2}g_3)$	$(f_3\lambda_{g_1}+\lambda_{f_3}g_1)$	$(f_3\lambda_{g_2}+\lambda_{f_3}g_2)$	$(f_3\lambda_{g_3}+\lambda_{f_3}g_3)$
$n \to pe\nu$	0.5072E-06	-0.3410E-09	0.6372E-13	0.3410E-09	-0.6243E-12	-0.4808E-16	-0.9251E-10	-0.4808E-16	-0.5043E-16
$\Sigma^+\to\Lambda e\nu$	0.7898E-03	-0.2843E-04	0.8640E-10	0.2843E-04	-0.1934E-05	-0.2624E-11	-0.2700E-08	-0.2624E-11	-0.2624E-11
$\Sigma^-\to\Lambda e\nu$	0.1587E-02	-0.6298E-04	0.1716E-09	0.6298E-04	-0.4711E-05	-0.5735E-11	-0.4855E-08	-0.5735E-11	-0.5735E-11
$\Lambda \to pe\nu$	0.3670E+00	-0.3470E-01	0.4682E-07	0.3470E-01	-0.5929E-02	-0.3640E-08	-0.5423E-06	-0.3640E-08	-0.3640E-08
$\Sigma^-\to ne\nu$	0.4044E+01	-0.5252E+00	0.4547E-06	0.5252E+00	-0.1201E+00	-0.4783E-07	-0.3770E-05	-0.4783E-07	-0.4783E-07
$\Xi^-\to\Lambda e\nu$	0.7434E+00	-0.6879E-01	0.6755E-07	0.6879E-01	-0.1152E-01	-0.5144E-08	-0.8002E-06	-0.5144E-08	-0.5144E-08
$\Xi^-\to\Sigma^o e\nu$	0.3037E-01	-0.1735E-02	0.2718E-08	0.1735E-02	-0.1842E-03	-0.1297E-09	-0.5301E-07	-0.1297E-09	-0.1297E-09
$\Sigma^o\to pe\nu$	0.3707E+01	-0.4765E+00	0.4200E-06	0.4765E+00	-0.1079E+00	-0.4375E-07	-0.3521E-05	-0.4375E-07	-0.4375E-07
$\Xi^o\to\Sigma^+ e\nu$	0.2562E-01	-0.1432E-02	0.2313E-08	0.1432E-02	-0.1489E-03	-0.1081E-09	-0.4614E-07	-0.1081E-09	-0.1081E-09
$\Xi^-\to\Xi^o e\nu$	0.2599E-10	-0.7118E-13	0.2227E-17	0.7118E-13	-0.3879E-15	-0.5305E-20	-0.9145E-15	-0.5305E-20	-0.5323E-20
$\Sigma^-\to\Sigma^o e\nu$	0.4638E-11	-0.1055E-13	0.4771E-18	0.1055E-13	-0.4857E-16	-0.9551E-21	-0.2336E-15	-0.9551E-21	-0.9607E-21
$\Sigma^o\to\Sigma^+ e\nu$	0.1929E-12	-0.2720E-15	0.1912E-19	0.2720E-15	-0.8182E-18	-0.2464E-22	-0.1469E-16	-0.2464E-22	-0.2498E-22
$\Xi^-\to ne\nu$	0.4663E+02	-0.8279E+01	0.4393E-05	0.8279E+01	-0.2501E+01	-0.6191E-06	-0.2602E-04	-0.6191E-06	-0.6191E-06
$\Xi^o\to pe\nu$	0.4299E+02	-0.7563E+01	0.4088E-05	0.7563E+01	-0.2266E+01	-0.5711E-06	-0.2445E-04	-0.5711E-06	-0.5711E-06
$\Sigma^+\to ne\nu$	0.3313E+01	-0.4192E+00	0.3770E-06	0.4192E+00	-0.9363E-01	-0.3869E-07	-0.3213E-05	-0.3869E-07	-0.3869E-07
$\Xi^o\to\Sigma^- e\nu$	0.1630E-01	-0.8518E-03	0.1470E-08	0.8518E-03	-0.8308E-04	-0.6432E-10	-0.3141E-07	-0.6432E-10	-0.6433E-10
$\Lambda \to p\mu\nu$	0.1170E+00	-0.8940E-02	0.3728E-03	0.8940E-02	-0.2121E-02	-0.3747E-04	-0.4318E-02	-0.3747E-04	-0.4010E-04
$\Sigma^-\to n\mu\nu$	0.2808E+01	-0.3080E+00	0.9580E-02	0.3080E+00	-0.8561E-01	-0.1138E-02	-0.7942E-01	-0.1138E-02	-0.1199E-02
$\Xi^-\to\Lambda\mu\nu$	0.3613E+00	-0.2744E-01	0.8922E-03	0.2744E-01	-0.6009E-02	-0.8241E-04	-0.1057E-01	-0.8241E-04	-0.8773E-04
$\Xi^-\to\Sigma^o\mu\nu$	0.9808E-03	-0.4528E-04	0.1802E-05	0.4528E-04	-0.7784E-05	-0.1362E-06	-0.3514E-04	-0.1362E-06	-0.1448E-06
$\Sigma^o\to p\mu\nu$	0.2534E+01	-0.2744E+00	0.8654E-02	0.2744E+00	-0.7588E-01	-0.1022E-02	-0.7255E-01	-0.1022E-02	-0.1077E-02
$\Xi^o\to\Sigma^+\mu\nu$	0.5525E-03	-0.2502E-04	0.1007E-05	0.2502E-04	-0.4275E-05	-0.7619E-07	-0.2008E-04	-0.7619E-07	-0.8077E-07
$\Xi^-\to n\mu\nu$	0.4188E+02	-0.6651E+01	0.1389E+00	0.6651E+01	-0.2232E+01	-0.2057E-01	-0.8228E+00	-0.2057E-01	-0.2126E-01
$\Xi^o\to p\mu\nu$	0.3843E+02	-0.6036E+01	0.1281E+00	0.6036E+01	-0.2014E+01	-0.1884E-01	-0.7665E+00	-0.1884E-01	-0.1949E-01
$\Sigma^+\to n\mu\nu$	0.2218E+01	-0.2359E+00	0.7547E-02	0.2359E+00	-0.6464E-01	-0.8821E-03	-0.6432E-01	-0.8821E-03	-0.9309E-03
$\Xi^o\to\Sigma^-\mu\nu$	0.8242E-04	-0.3522E-05	0.1435E-06	0.3522E-05	-0.5865E-06	-0.1083E-07	-0.3067E-05	-0.1083E-07	-0.1137E-07

Table 7(A)

Process	f_1^2	f_1f_2	f_2^2
$n \to pe\nu$	-0.1037E-03	-0.2073E-03	0.4727E-07
$\Sigma^+ \to \Lambda e\nu$	-0.7492E-02	-0.1443E-01	0.1813E-03
$\Sigma^- \to \Lambda e\nu$	-0.1369E-01	-0.2626E-01	0.3611E-03
$\Lambda \to pe\nu$	-0.1379E+01	-0.2500E+01	0.7223E-01
$\Sigma^- \to ne\nu$	-0.1133E+02	-0.1982E+02	0.7152E+00
$\Xi^- \to \Lambda e\nu$	-0.2851E+01	-0.5178E+01	0.1472E+00
$\Xi^- \to \Sigma^0 e\nu$	-0.1842E+00	-0.3470E+00	0.6617E-02
$\Sigma^0 \to pe\nu$	-0.1049E+02	-0.1837E+02	0.6585E+00
$\Xi^0 \to \Sigma^+ e\nu$	-0.1587E+00	-0.2992E+00	0.5598E-02
$\Xi^- \to \Xi^0 e\nu$	-0.2988E-08	-0.5959E-08	0.5996E-11
$\Sigma^- \to \Sigma^0 e\nu$	-0.6166E-09	-0.1230E-08	0.1021E-11
$\Sigma^0 \to \Sigma^+ e\nu$	-0.3650E-10	-0.7288E-10	0.3682E-13
$\Xi^- \to ne\nu$	-0.9870E+02	-0.1648E+03	0.6979E+01
$\Xi^0 \to pe\nu$	-0.9175E+02	-0.1534E+03	0.6475E+01
$\Sigma^+ \to ne\nu$	-0.9505E+01	-0.1668E+02	0.5922E+00
$\Xi^0 \to \Sigma^- e\nu$	-0.1077E+00	-0.2039E+00	0.3596E-02
$\Lambda \to p\mu\nu$	-0.1105E+00	-0.2014E+00	0.3734E-02
$\Sigma^- \to n\mu\nu$	-0.3604E+01	-0.6357E+01	0.1329E+00
$\Xi^- \to \Lambda\mu\nu$	-0.4626E+00	-0.8451E+00	0.1489E-01
$\Xi^- \to \Sigma^0\mu\nu$	-0.4927E-03	-0.9250E-03	0.2360E-04
$\Sigma^0 \to p\mu\nu$	-0.3228E+01	-0.5702E+01	0.1182E+00
$\Xi^0 \to \Sigma^+\mu\nu$	-0.2394E-03	-0.4494E-03	0.1271E-04
$\Xi^- \to n\mu\nu$	-0.5800E+02	-0.9772E+02	0.2610E+01
$\Xi^0 \to p\mu\nu$	-0.5316E+02	-0.8971E+02	0.2379E+01
$\Sigma^+ \to n\mu\nu$	-0.2806E+01	-0.4966E+01	0.1018E+00
$\Xi^0 \to \Sigma^-\mu\nu$	-0.2104E-04	-0.3935E-04	0.1624E-05

Table 7(B)

Process	$f_1\lambda_{f_1}$	$(f_1\lambda_{f_2}+\lambda_{f_1}f_2)$	$f_2\lambda_{f_2}$
$n \to pe\nu$	-0.2651E-09	-0.2649E-09	0.5492E-13
$\Sigma^+ \to \Lambda e\nu$	-0.2191E-04	-0.2114E-04	0.3535E-06
$\Sigma^- \to \Lambda e\nu$	-0.4868E-04	-0.4680E-04	0.8525E-06
$\Lambda \to pe\nu$	-0.2811E-01	-0.2563E-01	0.9063E-03
$\Sigma^- \to ne\nu$	-0.4401E+00	-0.3883E+00	0.1601E-01
$\Xi^- \to \Lambda e\nu$	-0.5562E-01	-0.5081E-01	0.1774E-02
$\Xi^- \to \Sigma^0 e\nu$	-0.1360E-02	-0.1285E-02	0.3175E-04
$\Sigma^0 \to pe\nu$	-0.3988E+00	-0.3522E+00	0.1447E-01
$\Xi^0 \to \Sigma^+ e\nu$	-0.1121E-02	-0.1061E-02	0.2575E-04
$\Xi^- \to \Xi^0 e\nu$	-0.5447E-13	-0.5432E-13	0.7528E-16
$\Sigma^- \to \Sigma^0 e\nu$	-0.8153E-14	-0.8134E-14	0.9265E-17
$\Sigma^0 \to \Sigma^+ e\nu$	-0.2154E-15	-0.2151E-15	0.1459E-18
$\Xi^- \to ne\nu$	-0.7302E+01	-0.6171E+01	0.2613E+00
$\Xi^0 \to pe\nu$	-0.6658E+01	-0.5635E+01	0.2392E+00
$\Sigma^+ \to ne\nu$	-0.3501E+00	-0.3098E+00	0.1266E-01
$\Xi^0 \to \Sigma^- e\nu$	-0.6652E-03	-0.6317E-03	0.1452E-04
$\Lambda \to p\mu\nu$	-0.5914E-02	-0.5465E-02	-0.7787E-04
$\Sigma^- \to n\mu\nu$	-0.2645E+00	-0.2357E+00	0.8838E-03
$\Xi^- \to \Lambda\mu\nu$	-0.2055E-01	-0.1896E-01	-0.5706E-04
$\Xi^- \to \Sigma^0\mu\nu$	-0.1488E-04	-0.1427E-04	-0.4222E-06
$\Sigma^0 \to p\mu\nu$	-0.2344E+00	-0.2092E+00	0.6590E-03
$\Xi^0 \to \Sigma^+\mu\nu$	-0.7347E-05	-0.7056E-05	-0.2284E-06
$\Xi^- \to n\mu\nu$	-0.6309E+01	-0.5373E+01	0.8431E-01
$\Xi^0 \to p\mu\nu$	-0.5713E+01	-0.4873E+01	0.7505E-01
$\Sigma^+ \to n\mu\nu$	-0.2000E+00	-0.1788E+00	0.4399E-03
$\Xi^0 \to \Sigma^-\mu\nu$	-0.6988E-06	-0.6748E-06	-0.2773E-07

Table 7(C)

Process	g_1^2	g_1g_2	g_2^2
$n\to pe\nu$	0.5059E+00	-0.1186E-02	0.6734E-06
$\Sigma^+\to\Lambda e\nu$	0.5878E+00	-0.5830E-01	0.1355E-02
$\Sigma^-\to\Lambda e\nu$	0.9729E+00	-0.1062E+00	0.2716E-02
$\Lambda\to pe\nu$	0.4027E+02	-0.1022E+02	0.6075E+00
$\Sigma^-\to ne\nu$	0.2368E+03	-0.8134E+02	0.6536E+01
$\Xi^-\to\Lambda e\nu$	0.8511E+02	-0.2116E+02	0.1232E+01
$\Xi^-\to\Sigma^0 e\nu$	0.9040E+01	-0.1409E+01	0.5146E-01
$\Sigma^0\to pe\nu$	0.2216E+03	-0.7538E+02	0.5998E+01
$\Xi^0\to\Sigma^+ e\nu$	0.7960E+01	-0.1215E+01	0.4343E-01
$\Xi^-\to\Xi^0 e\nu$	0.3108E-05	-0.2424E-07	0.4440E-10
$\Sigma^-\to\Sigma^0 e\nu$	0.7721E-06	-0.5063E-08	0.7808E-11
$\Sigma^0\to\Sigma^+ e\nu$	0.7405E-07	-0.3121E-09	0.3109E-12
$\Xi^-\to ne\nu$	0.1470E+04	-0.6767E+03	0.7277E+02
$\Xi^0\to pe\nu$	0.1381E+04	-0.6301E+03	0.6718E+02
$\Sigma^+\to ne\nu$	0.2043E+03	-0.6846E+02	0.5367E+01
$\Xi^0\to\Sigma^- e\nu$	0.5788E+01	-0.8273E+00	0.2770E-01
$\Lambda\to p\mu\nu$	0.6390E+01	-0.1826E+01	0.1288E+00
$\Sigma^-\to n\mu\nu$	0.1034E+03	-0.3803E+02	0.3394E+01
$\Xi^-\to\Lambda\mu\nu$	0.2274E+02	-0.6220E+01	0.4171E+00
$\Xi^-\to\Sigma^0\mu\nu$	0.1160E+00	-0.2167E-01	0.1011E-02
$\Sigma^0\to p\mu\nu$	0.9454E+02	-0.3447E+02	0.3052E+01
$\Xi^0\to\Sigma^+\mu\nu$	0.6698E-01	-0.1233E-01	0.5663E-03
$\Xi^-\to n\mu\nu$	0.1002E+04	-0.4781E+03	0.5450E+02
$\Xi^0\to p\mu\nu$	0.9313E+03	-0.4410E+03	0.4991E+02
$\Sigma^+\to n\mu\nu$	0.8451E+02	-0.3042E+02	0.2661E+01
$\Xi^0\to\Sigma^-\mu\nu$	0.1091E-01	-0.1910E-02	0.8353E-04

Table 7(D)

Process	$g_1\lambda_{g_1}$	$(g_1\lambda_{g_2}+\lambda_{g_1}g_2)$	$g_2\lambda_{g_2}$
$n\to pe\nu$	0.1015E-05	-0.1132E-08	0.1194E-11
$\Sigma^+\to\Lambda e\nu$	0.1602E-02	-0.7800E-04	0.3515E-05
$\Sigma^-\to\Lambda e\nu$	0.3223E-02	-0.1728E-03	0.8569E-05
$\Lambda\to pe\nu$	0.7622E+00	-0.9503E-01	0.1095E-01
$\Sigma^-\to ne\nu$	0.8528E+01	-0.1439E+01	0.2242E+00
$\Xi^-\to\Lambda e\nu$	0.1543E+01	-0.1884E+00	0.2127E-01
$\Xi^-\to\Sigma^0 e\nu$	0.6211E-01	-0.4755E-02	0.3367E-03
$\Sigma^0\to pe\nu$	0.7814E+01	-0.1305E+01	0.2014E+00
$\Xi^0\to\Sigma^+ e\nu$	0.5235E-01	-0.3924E-02	0.2721E-03
$\Xi^-\to\Xi^0 e\nu$	0.5204E-10	-0.1985E-12	0.7004E-15
$\Sigma^-\to\Sigma^0 e\nu$	0.9284E-11	-0.2970E-13	0.8788E-16
$\Sigma^0\to\Sigma^+ e\nu$	0.3861E-12	-0.7885E-15	0.1491E-17
$\Xi^-\to ne\nu$	0.1006E+03	-0.2273E+02	0.4740E+01
$\Xi^0\to pe\nu$	0.9264E+02	-0.2076E+02	0.4293E+01
$\Sigma^+\to ne\nu$	0.6976E+01	-0.1148E+01	0.1746E+00
$\Xi^0\to\Sigma^- e\nu$	0.3327E-01	-0.2335E-02	0.1516E-03
$\Lambda\to p\mu\nu$	0.2400E+00	-0.3273E-01	0.4340E-02
$\Sigma^-\to n\mu\nu$	0.5881E+01	-0.1030E+01	0.1708E+00
$\Xi^-\to\Lambda\mu\nu$	0.7431E+00	-0.9676E-01	0.1212E-01
$\Xi^-\to\Sigma^0\mu\nu$	0.1976E-02	-0.1787E-03	0.1606E-04
$\Sigma^0\to p\mu\nu$	0.5302E+01	-0.9205E+00	0.1515E+00
$\Xi^0\to\Sigma^+\mu\nu$	0.1112E-02	-0.9927E-04	0.8816E-05
$\Xi^-\to n\mu\nu$	0.9006E+02	-0.2060E+02	0.4385E+01
$\Xi^0\to p\mu\nu$	0.8257E+02	-0.1873E+02	0.3957E+01
$\Sigma^+\to n\mu\nu$	0.4635E+01	-0.7944E+00	0.1292E+00
$\Xi^0\to\Sigma^-\mu\nu$	0.1655E-03	-0.1414E-04	0.1205E-05

Table 7(E)

Process	f_1g_1	f_1g_2	f_1g_3	f_2g_1	f_2g_2	f_2g_3	f_3g_1	f_3g_2	f_3g_3
$n \to pe\nu$	0.1542E-04	0.7455E-04	0.1291E-07	0.2853E-03	-0.3417E-06	0.1510E-10	0.8999E-04	-0.1497E-06	-0.1559E-10
$\Sigma^+\to\Lambda e\nu$	0.9063E-06	0.1290E-05	-0.1303E-07	0.7487E-05	-0.3232E-06	0.1383E-08	0.2179E-05	-0.1085E-06	-0.2860E-12
$\Sigma^-\to\Lambda e\nu$	0.1357E-05	0.1904E-05	-0.2121E-07	0.1113E-04	-0.5284E-06	0.2493E-08	0.3232E-05	-0.1772E-06	-0.4199E-12
$\Lambda \to pe\nu$	0.3119E-04	0.3675E-04	-0.9495E-06	0.2330E-03	-0.2545E-04	0.2894E-06	0.6597E-04	-0.8449E-05	-0.1029E-10
$\Sigma^-\to ne\nu$	0.1265E-03	0.1338E-03	-0.4593E-05	0.8920E-03	-0.1307E-03	0.2063E-05	0.2480E-03	-0.4313E-04	-0.3452E-10
$\Xi^-\to\Lambda e\nu$	0.4786E-04	0.5656E-04	-0.1435E-05	0.3582E-03	-0.3835E-04	0.4264E-06	0.1015E-03	-0.1273E-04	-0.1128E-10
$\Xi^-\to\Sigma^o e\nu$	0.7550E-05	0.9950E-05	-0.1592E-06	0.5987E-04	-0.4050E-05	0.2755E-07	0.1725E-04	-0.1352E-05	-0.1867E-11
$\Sigma^o\to pe\nu$	0.1202E-03	0.1277E-03	-0.4344E-05	0.8495E-03	-0.1233E-03	0.1926E-05	0.2364E-03	-0.4070E-04	-0.3314E-10
$\Xi^0\to\Sigma^+ e\nu$	0.6839E-05	0.9050E-05	-0.1417E-06	0.5435E-04	-0.3600E-05	0.2396E-07	0.1567E-04	-0.1202E-05	-0.1710E-11
$\Xi^-\to\Xi^0 e\nu$	0.3559E-10	0.7737E-10	-0.4175E-13	0.3745E-09	-0.1274E-11	0.4358E-15	0.1129E-09	-0.4649E-12	-0.1111E-16
$\Sigma^-\to\Sigma^0 e\nu$	0.1174E-10	0.2791E-10	-0.1043E-13	0.1307E-09	-0.3766E-12	0.1076E-15	0.3964E-10	-0.1406E-12	-0.4681E-17
$\Sigma^0\to\Sigma^+ e\nu$	0.1455E-11	0.4154E-11	-0.4887E-15	0.1828E-10	-0.3474E-13	0.6054E-17	0.5609E-11	-0.1360E-13	-0.6486E-18
$\Xi^-\to ne\nu$	0.5320E-03	0.4864E-03	-0.2134E-04	0.3470E-02	-0.6733E-03	0.1476E-04	0.9387E-03	-0.2199E-03	-0.1116E-09
$\Xi^0\to pe\nu$	0.5071E-03	0.4659E-03	-0.2031E-04	0.3317E-02	-0.6383E-03	0.1386E-04	0.8980E-03	-0.2086E-03	-0.1076E-09
$\Sigma^+\to ne\nu$	0.1126E-03	0.1203E-03	-0.4038E-05	0.7981E-03	-0.1142E-03	0.1755E-05	0.2223E-03	-0.3771E-04	-0.3128E-10
$\Xi^0\to\Sigma^- e\nu$	0.5269E-05	0.7057E-05	-0.1035E-06	0.4214E-04	-0.2616E-05	0.1627E-07	0.1217E-04	-0.8742E-06	-0.1325E-11
$\Lambda \to p\mu\nu$	0.3454E-01	0.2120E+00	0.9691E-02	0.7709E+00	-0.1228E+00	-0.2622E-03	0.2546E+00	-0.5857E-01	-0.1253E-02
$\Sigma^-\to n\mu\nu$	0.6352E+00	0.2453E+01	0.9061E-01	0.9785E+01	-0.1799E+01	0.1379E-01	0.3121E+01	-0.8198E+00	-0.1427E-01
$\Xi^-\to\Lambda\mu\nu$	0.1119E+00	0.5777E+00	0.2092E-01	0.2170E+01	-0.3170E+00	0.4126E-03	0.7052E+00	-0.1480E+00	-0.2546E-02
$\Xi^-\to\Sigma^o\mu\nu$	0.3328E-03	0.3984E-02	0.1622E-03	0.1327E-01	-0.1539E-02	-0.1170E-04	0.4474E-02	-0.7565E-03	-0.1485E-04
$\Sigma^o\to p\mu\nu$	0.5803E+00	0.2275E+01	0.8500E-01	0.9043E+01	-0.1655E+01	0.1206E-01	0.2889E+01	-0.7555E+00	-0.1328E-01
$\Xi^o\to\Sigma^+\mu\nu$	0.1804E-03	0.2349E-02	0.9636E-04	0.7764E-02	-0.8947E-03	-0.7195E-05	0.2624E-02	-0.4412E-03	-0.8742E-05
$\Xi^-\to n\mu\nu$	0.6007E+01	0.1564E+02	0.3993E+00	0.6904E+02	-0.1504E+02	0.2855E+00	0.2109E+02	-0.6491E+01	-0.8536E-01
$\Xi^o\to p\mu\nu$	0.5608E+01	0.1478E+02	0.3841E+00	0.6502E+02	-0.1409E+02	0.2622E+00	0.1990E+02	-0.6095E+01	-0.8110E-01
$\Sigma^+\to n\mu\nu$	0.5169E+00	0.2066E+01	0.7803E-01	0.8176E+01	-0.1484E+01	0.1009E-01	0.2616E+01	-0.6790E+00	-0.1205E-01
$\Xi^0\to\Sigma^-\mu\nu$	0.2321E-04	0.3989E-03	0.1643E-04	0.1289E-02	-0.1446E-03	-0.1301E-05	0.4383E-03	-0.7188E-04	-0.1437E-05

Table 7(F)

Process	$(f_1\lambda_{g_1}+\lambda_{f_1}g_1)$	$(f_1\lambda_{g_2}+\lambda_{f_1}g_2)$	$(f_1\lambda_{g_3}+\lambda_{f_1}g_3)$	$(f_2\lambda_{g_1}+\lambda_{f_2}g_1)$	$(f_2\lambda_{g_2}+\lambda_{f_2}g_2)$	$(f_2\lambda_{g_3}+\lambda_{f_2}g_3)$	$(f_3\lambda_{g_1}+\lambda_{f_3}g_1)$	$(f_3\lambda_{g_2}+\lambda_{f_3}g_2)$	$(f_3\lambda_{g_3}+\lambda_{f_3}g_3)$
$n\to pe\nu$	-0.1592E-10	0.7915E-10	0.1134E-13	0.2325E-09	-0.2563E-12	0.2787E-16	0.9259E-10	-0.1501E-12	-0.1133E-16
$\Sigma^+\to\Lambda e\nu$	-0.8181E-09	0.1234E-09	-0.1148E-10	0.4060E-08	-0.1933E-09	0.2022E-11	0.2775E-08	-0.1478E-09	-0.1817E-15
$\Sigma^-\to\Lambda e\nu$	-0.1480E-08	0.2231E-09	-0.2263E-10	0.7316E-08	-0.3829E-09	0.4433E-11	0.5004E-08	-0.2935E-09	-0.3230E-15
$\Lambda\to pe\nu$	-0.1815E-06	0.3214E-07	-0.5369E-08	0.8475E-06	-0.1009E-06	0.2949E-08	0.5838E-06	-0.7996E-07	-0.4277E-13
$\Sigma^-\to ne\nu$	-0.1341E-05	0.2746E-06	-0.4632E-07	0.6041E-05	-0.9580E-06	0.4008E-07	0.4178E-05	-0.7767E-06	-0.2626E-12
$\Xi^-\to\Lambda e\nu$	-0.2668E-06	0.4672E-07	-0.7792E-08	0.1248E-05	-0.1458E-06	0.4159E-08	0.8599E-06	-0.1154E-06	-0.4493E-13
$\Xi^-\to\Sigma^0 e\nu$	-0.1665E-07	0.2574E-08	-0.3444E-09	0.8074E-07	-0.5997E-08	0.1017E-09	0.5539E-07	-0.4645E-08	-0.2923E-14
$\Sigma^0\to pe\nu$	-0.1250E-05	0.2544E-06	-0.4301E-07	0.5637E-05	-0.8857E-06	0.3661E-07	0.3898E-05	-0.7174E-06	-0.2471E-12
$\Xi^0\to\Sigma^+ e\nu$	-0.1446E-07	0.2229E-08	-0.2940E-09	0.7022E-07	-0.5109E-08	0.8467E-10	0.4816E-07	-0.3954E-08	-0.2567E-14
$\Xi^-\to\Xi^0 e\nu$	-0.2497E-15	0.1830E-15	-0.2790E-18	0.1517E-14	-0.5355E-17	0.4023E-20	0.9164E-15	-0.3892E-17	-0.4988E-22
$\Sigma^-\to\Sigma^0 e\nu$	-0.6202E-16	0.6330E-16	-0.5396E-19	0.4075E-15	-0.1185E-17	0.7268E-21	0.2341E-15	-0.8455E-18	-0.1567E-22
$\Sigma^0\to\Sigma^+ e\nu$	-0.3605E-17	0.6385E-17	-0.1384E-20	0.2859E-16	-0.5214E-19	0.1876E-22	0.1471E-16	-0.3545E-19	-0.1019E-23
$\Xi^-\to ne\nu$	-0.1007E-04	0.2544E-05	-0.3601E-06	0.4322E-04	-0.8969E-05	0.5461E-06	0.3005E-04	-0.7520E-05	-0.1521E-11
$\Xi^o\to pe\nu$	-0.9438E-05	0.2366E-05	-0.3381E-06	0.4057E-04	-0.8354E-05	0.5028E-06	0.2820E-04	-0.6996E-05	-0.1443E-11
$\Sigma^+\to ne\nu$	-0.1136E-05	0.2294E-06	-0.3889E-07	0.5137E-05	-0.7960E-06	0.3232E-07	0.3551E-05	-0.6439E-06	-0.2265E-12
$\Xi^o\to\Sigma^- e\nu$	-0.9782E-08	0.1494E-08	-0.1884E-09	0.4769E-07	-0.3253E-08	0.5023E-10	0.3269E-07	-0.2512E-08	-0.1744E-14
$\Lambda\to p\mu\nu$	-0.5852E-03	0.4784E-02	0.1787E-03	0.1299E-01	-0.1929E-02	0.8110E-05	0.4869E-02	-0.1095E-02	-0.1841E-04
$\Sigma^-\to n\mu\nu$	-0.1570E-01	0.7995E-01	0.2383E-02	0.2266E+00	-0.3770E-01	0.7443E-03	0.9139E-01	-0.2318E-01	-0.2854E-03
$\Xi^-\to\Lambda\mu\nu$	-0.1664E-02	0.1118E-01	0.3286E-03	0.3077E-01	-0.4143E-02	0.3437E-04	0.1179E-01	-0.2407E-02	-0.3132E-04
$\Xi^-\to\Sigma^0\mu\nu$	-0.2679E-05	0.3901E-04	0.1382E-05	0.1075E-03	-0.1205E-04	-0.6198E-07	0.3833E-04	-0.6429E-05	-0.1096E-06
$\Sigma^0\to p\mu\nu$	-0.1418E-01	0.7339E-01	0.2211E-02	0.2074E+00	-0.3438E-01	0.6569E-03	0.8341E-01	-0.2107E-01	-0.2631E-03
$\Xi^0\to\Sigma^+\mu\nu$	-0.1420E-05	0.2224E-04	0.8008E-06	0.6158E-04	-0.6877E-05	-0.3978E-07	0.2189E-04	-0.3655E-05	-0.6350E-07
$\Xi^-\to n\mu\nu$	-0.2182E+00	0.7067E+00	0.1513E-01	0.2194E+01	-0.4277E+00	0.1779E-01	0.9769E+00	-0.2903E+00	-0.2377E-02
$\Xi^o\to p\mu\nu$	-0.2015E+00	0.6625E+00	0.1440E-01	0.2049E+01	-0.3974E+00	0.1621E-01	0.9090E+00	-0.2688E+00	-0.2238E-02
$\Sigma^+\to n\mu\nu$	-0.1238E-01	0.6544E-01	0.1993E-02	0.1844E+00	-0.3035E-01	0.5543E-03	0.7386E-01	-0.1853E-01	-0.2348E-03
$\Xi^0\to\Sigma^-\mu\nu$	-0.1687E-06	0.3364E-05	0.1248E-06	0.9447E-05	-0.1036E-05	-0.7564E-08	0.3335E-05	-0.5443E-06	-0.9820E-08

Addendum

by A. Bohm

After these lecture notes were completed and sent to the publisher, a new experimental value for the electron asymmetry in $\Sigma^- \to ne\nu$ was published.[1)] This value is based on approximately 25,000 decays in Fermilab experiment 715 with a Σ^- beam of approximately 20% polarization (compared to a total of 193 previous events). As this experiment is far superior to the previous experiments, its value will dominate the world average. The preliminary result is $\alpha_e^{\Sigma^- \to n} = -0.58 \pm 0.16$ and is in contradiction to the present world average of $\alpha_e^{\Sigma \to n} = +0.26 \pm 0.19$ given in Table 4 of Chapter 3. It is in agreement with the prediction of the Cabibbo theory which -- with this new value -- is in very good agreement with all hyperon semileptonic decay data.

The Cabibbo model as described in Chapter 4 does not take the mass differences of the hyperons into account, except in the phase space factors. One may, therefore, wonder why the Cabibbo model without symmetry breaking gives such an accurate description of the experimental situation. The mass differences are taken into account in the model in which SU(3) is treated as a spectrum generating group, the SG model of chapter 9. The formfactors $f_i^{\alpha'\beta\alpha}$ and $g_i^{\alpha'\beta\alpha}$ for each individual process hyperon $\alpha \to$ hyperon $\alpha' + \ell\nu$ are, for the uncorrected Cabibbo model, expressed in terms of the SU(3) octet formfactors $F_i^{(1)} = F_i$ $F_i^{(2)} = D_i$ $G_1^{(1)} = F$ $G_1^{(2)} = D$ by equation (4) and (5) of chapter 4. For the SG model these formfactors are expressed in terms of the SU(3) octet form factors by formulas (11) of chapter 9 with $G_2^{(\gamma)} = 0$ and $F_3^{(\gamma)} = 0$. With $F_i^{(\gamma)}$ determined by CVC one has therefore in the uncorrected Cabibbo model the three free parameters $\sin\theta$ F, D whereas in the SG model one has the five parameters

$\sin\theta$, F, D, $G_3^{(F)}$, $G_3^{(D)}$. There is a priori no reason for which $G_3^{F,D}$ should be zero in the Cabibbo model nor in the SG model. But in the Cabibbo model $G_3^{F,D}$ only contribute to $g_3^{\alpha'\beta\alpha}$ which enters the experimental quantities with a factor of m_e, whereas in the SG model $G_3^{F,D}$ can also contribute to $g_2^{\alpha'\beta\alpha}$ and could therefore have an observable effect.

The expressions for the formfactors in the SG model go into the expressions for the uncorrected Cabibbo model if the hyperon mass differences are zero. But the axial formfactors $g_i^{\alpha'\beta\alpha}$ in the SG model are also identical to the axial formfactors in the Cabibbo model if $G_3^F = G_3^D = 0$. The vector formfactors $f_1^{\alpha'\beta\alpha}$ of both models are approximately equal. Thus the hyperon-semileptonic-decay data cannot distinguish between the Cabibbo model and the SG model if $G_3^{F,D} = 0$.

If one does not include $\alpha_e^{\Sigma\to n}$ in the experimental data to be fitted, then the five parameters of the SG model are not uniquely defined. There are two sets of values for F, D, G_3^F, G_3^D with approximately the same value for $\sin\theta$, which give an equally good fit to the set of all experimental data with $\alpha_e^{\Sigma-n}$ excluded. One of these fits gives large negative values for G_3^F and G_3^D. The other gives $G_3^F \approx 0$, $G_3^D \approx 0$. If one includes the old world average, $\alpha_e^{\Sigma\to n} = 0.26 \pm 0.19$, the fit will choose the solution with large G_3. If one includes the new value, $\alpha_e^{\Sigma\to n} = -0.58 \div 0.16$, the fit will prefer the solution with zero G_3. Only the first solution has been discussed in chapter 9.

As the experimental situation has changed we give in this Addendum a fit of the SG model to the new experimental data. We also give a fit of the uncorrected Cabibbo model. Table A presents the comparison between the latest values (October 1984) of 25 experimental quantities and the prediction of the Cabibbo model (given in Column 3) without any symmetry breaking. The agreement is truly remarkable. The only significant deviation, which occurs in $R(\Sigma^- \to \Lambda e\nu)$, can be easily explained by small perturbative corrections as discussed in section 8.1 or section 8.3. There are also small deviations for the asymmetries in $\Lambda \to pe\nu$, which are not yet statistically significant. If they persist they will be an indication that leptonic currents cannot be pure V-A.

Column 5 gives the predictions for the SG model in which the parameters have been determined from the experimental data in column 2. We see that the fits in column 3 and in column 5 are indistinguishable. Their total χ^2 values are also comparable. Table B gives the values of the parameters and we see that the values of $G_3^{F,D}$ are compatible with zero. If one fixes $G_3^{F,D}$ at zero, the χ^2 will not change and the fitted values will remain essentially the same.

A χ^2 of about 40 for about 20 degrees of freedom may not look so good. However, if one excludes $R(\Sigma^- \to \Lambda e\nu)$ from the fit, the total χ^2 will go down to about 20. And if one includes a correction term of about 10% for symmetry breaking, as explained in section 8.3 and 9.4, the χ^2 will then go down to about 15.

Thus we conclude that with the new value for $\alpha_e^{\Sigma\to n}$ there is an unbelievably good agreement between the experimental data and the predictions of the Cabibbo model. The three parameters of the Cabibbo model are already determined by the $n \to pe\nu$ data and by the rate and $\alpha_{e\nu}$ for $\Lambda \to pe\nu$, so that $\alpha_e^{\Sigma\to n} \approx -0.6$ and all the other values in column 3 can already be predicted by the model from this minimal set of data. The SG model gives essentially the same predictions but requires the experimental value for $\alpha_e^{\Sigma\to n}$ to determine the two additional parameters G_3^F and G_3^D.

1) Electron Asymmetry from Σ^- Beta Decay, Elmhurst College, Fermi Lab, U. of Iowa, Iowa State U., Leningrad Nuclear Phys. Inst., U. of Chicago. DPF Santa Fe Meeting 1984.

Table A

	Experimental Value	I Predicted Values Standard Model	Contribution to χ^2	II Predicted Values Model with Mass-corrections	
$n \to pe\nu R$	1.114 ± 0.020	1.095	0.9	1.085	0.9
$\alpha_{e\nu}$	-0.074 ± 0.004	-0.074	0.0	-0.074	0.2
α_e	-0.083 0.002	-0.082	0.5	-0.081	0.0
α_ν	0.998 0.025	0.989	0.1	0.989	0.1
α_p		-0.48		-0.43	
g_1/f_1	1.254 ± 0.006	1.249		1.249	
$\Sigma^+ \to \Lambda e\nu R$	0.250 ± 0.063	0.276	0.2	0.276	0.2
$\alpha_{e\nu}$	-0.35 ± 0.15	-0.41	0.1	-0.40	0.1
f_1/g_1	-0.37 ± 0.22	±0.00		-0.004	
$\Sigma^- \to \Lambda e\nu R$	0.387 ± 0.018	0.458	15.7	0.456	14.9
$\alpha_{e\nu}$	-0.404 ± 0.044	-0.412	0.0	-0.408	0.0
A	0.07 ± 0.07	0.06	0.0	0.04	0.1
B	0.85 ± 0.07	0.88	0.2	0.88	0.2
f_1/g_1	-0.14 ± 0.24	±0.000		-0.004	
$\Lambda \to pe\nu R$	3.180 ± 0.058	3.207	0.2	3.239	1.0
$\alpha_{e\nu}$	-0.013 ± 0.014	-0.019	0.2	-0.025	0.8
α_e	0.125 ± 0.066	0.009	3.1	0.007	3.2
α_ν	0.821 ± 0.060	0.977	6.7	0.984	7.4
α_n	-0.508 ± 0.065	-0.578	1.1	-0.582	1.3
f_1/g_1	0.719 ± 0.023	0.717		0.759	
$\Sigma^- \to ne\nu R$	6.896 ± 0.235	6.768	0.3	6.550	0.2
$\alpha_{e\nu}$	0.279 ± 0.026	0.333	4.2	0.296	0.4
α_e	+0.58 ± 0.16	-0.618	0.0	-0.671	0.3
α_ν		-0.389		-0.386	
α_n		0.694		0.726	
g_1/f_1		-0.349		-0.391	
$\Xi^- \to \Lambda e\nu R$	3.352 ± 0.367	2.876	1.6	2.723	2.9
$\alpha_{e\nu}$	0.53 ± 0.10	0.654	1.5	0.664	1.7
A	0.62 ± 0.1	0.455	2.7	0.448	2.9
g_1/f_1	0.248 ± 0.05	0.184		0.182	
$\Xi^- \to \Lambda^0 e\nu R$	0.53 ± 0.10	0.51	0.0	0.55	0.0
g_1/f_1		1.25		1.29	
$\Lambda \to p\mu\nu R$	0.596 ± 0.133	0.549	1.4	0.550	0.1
$\Sigma^- \to n\mu\nu R$	3.036 ± 0.271	3.158	2.0	3.008	0.0
$\Xi^- \to \Lambda\mu\nu R$	2.133 ± 2.133	0.819	0.4	0.775	0.4
χ^2			39		41

Comparison between the experimental hyperon-semileptonic-decay data and the predictions of the Cabibbo model. Only those data in column 2 for which we give a contribution to χ^2 in columns 4 and 6 have been used in the fit for the determination of the parameters. The g_1/f_1 ratios in columns 3 and 5 have been calculated from the parameters in Table B. We also list the predictions for the other asymmetries in $\Sigma^- \to ne\nu$.

Table B. Values of the Parameters

	Fit I	Fit II
$\sin\theta$	0.225 ± 0.002	0.239 ± 0.003
$1/\sqrt{6}\ G_1^F = F$	0.450 ± 0.006	0.440 ± 0.006
$-\sqrt{3/10}\ G_1^D = D$	0.799 ± 0.007	0.809 ± 0.007
$1/\sqrt{6}\ G_3^F$		-0.92 ± 1.28
$-\sqrt{3/10}\ G_3^D$		-1.38 ± 2.32

Values of the parameters. The values for $G_3^{F,D}$ are consistent with zero. In a fit in which they are fixed at zero the predictions in column 5 of Table A remain the same (except for minor changes in the third decimal). However, $G_3^{F,D}$ could also be of order one.